AF626069

Mathematical Research

Computer Analysis of Images and Patterns

edited by
L. P. Yaroslavskiĭ · A. Rosenfeld
W. Wilhelmi

Band 40

AKADEMIE-VERLAG BERLIN

In this series original contributions of mathematical research in all fields are contained, such as
- research monographs
- collections of papers to a single topic
- reports on congresses of exceptional interest for mathematical research.

This series is aimed at promoting quick information and communication between mathematicians of the various special branches.

In diese Reihe werden Originalbeiträge zu allen Gebieten der mathematischen Forschung aufgenommen wie
- Forschungsmonographien
- Sammlungen von Arbeiten zu einem speziellen Thema
- Berichte von Tagungen, die für die mathematische Forschung besonders aktuell sind.

Die Reihe soll die schnelle Information und gute Kommunikation zwischen den Mathematikern der verschiedenen Fachgebiete fördern.

Manuscripts
in English and German comprising at least 100 pages and not more than 500 pages can be admitted to this series. With respect to a quick publication the manuscripts are reproduced photomechanically. Authors who are interested in this series please turn directly to the 'Akademie-Verlag'. Here you will get more detailed information about the form of the manuscripts and the modalities of publication.

Manuskripte
in englischer und deutscher Sprache, die mindestens 100 Seiten und nicht mehr als 500 Seiten umfassen, können in diese Reihe aufgenommen werden. Im Interesse einer schnellen Publikation werden die Manuskripte auf fotomechanischem Weg reproduziert. Autoren, die an der Veröffentlichung entsprechender Arbeiten in dieser Reihe interessiert sind, wenden sich bitte direkt an den Akademie-Verlag. Sie erhalten dort genauere Informationen über die Gestaltung der Manuskripte und die Modalitäten der Veröffentlichung.

Computer Analysis of Images and Patterns

Mathematical Research · Mathematische Forschung

Wissenschaftliche Beiträge
herausgegeben von der
Akademie der Wissenschaften der DDR
Karl-Weierstraß-Institut für Mathematik

Band 40

Computer Analysis of
Images and Patterns

Computer Analysis of Images and Patterns

Proceedings of the II. International Conference CAIP '87 on Automatic Image Processing held in Wismar (GDR), September 2–4, 1987

edited by

Leonid P. Yaroslavskiĭ

Azriel Rosenfeld,

Wolfgang Wilhelmi

Akademie-Verlag Berlin 1987

Herausgeber:
Prof. Dr. sc.techn. Leonid P. Yaroslavskiĭ
Akademie der Wissenschaften der UdSSR
Institut für Probleme der Informationsübertragung (IPPI)

Prof. Dr. phil. Azriel Rosenfeld
University of Maryland
Center for Automation Research

Prof. Dr. sc.techn. Wolfgang Wilhelmi
Akademie der Wissenschaften der DDR
Zentralinstitut für Kybernetik und Informationsprozesse

Die Titel dieser Schriftenreihe werden vom Originalmanuskript der Autoren reproduziert

ISBN 3-05-500451-5
ISSN 0138-3019

Erschienen im Akademie-Verlag Berlin,DDR-1086 Berlin, Leipziger Str.3-4
© Akademie-Verlag Berlin 1987
Lizenznummer: 202 · 100/410/87
Printed in the German Democratic Republic
Gesamtherstellung: VEB Kongreß- und Werbedruck, 9273 Oberlungwitz
Lektor: Dipl.-Phys. Gisela Lagowitz
LSV 1085
Bestellnummer: 763 790 6 (2182/40)

03800

Foreword

This volume is a collection of selected papers delivered at the II. International Conference on Automatic Image Processing - CAIP '87 - held in Wismar, GDR from September 2 to 4, 1987.
The success achieved by the I. International Conference on Automatic Image Processing (October 17-18,1985 in Berlin) has encouraged the Image Processing Group in the Scientific-Technological Society for Measurement and Automatic Control (WGMA) to organize a second conference two years later. In doing so we hope that CAIP will increase its importance and its role so far played among expert circles.
The presented contributions have shown the great influence reached by image processing and image recognition methods on the introduction of new kinds of information-processing principles (especially parallel ones) and the progress towards new generations of machines capable of understanding real world scenes. On the other hand, CAIP '87 demonstrated the penetration of artificial intelligence methods into the image analysis field. Undoubtedly reader will not observe spectacular breakthroughs but rather a well balanced continuous advance.
The quality of papers accepted for publication was essentially dependent on the quality of refereeing. Each paper except for those invited was evaluated by two members of the International Programm Committee. The final decision was taken by W.Wilhelmi and R.Klette in cooperation with the Executive Board of Image Processing Group in WGMA. Contributions showing the highest score in scientific level, innovative approach and topicality to CAIP were selected for publication in this volume. A complete overview including poster presentations is given by another publication by the Kammer der Technik (KdT) containing 2 pages of abstract on every contribution.

The editors would like to express their thanks to the other members of the program committee

D.Chetverikov (Budapest)

S.Fuchs (Dresden)

B.Gudmundsson (Linköping)

R.Klette (Berlin)

V.A.Kovalevsky (Berlin)

C.-E.Liedtke (Hannover)

M.I.Schlesinger (Kiev)

D.Schmidt (Wismar)

F.Sloboda (Bratislava)

K.Voss (Jena)

for their competent and reliable cooperation, to the invited speakers and chairman of the sessions, and to all other contributors. A great help for the editors was the support of Dr. R.Klette during the evaluation and decision making process. We are very much obliged to Prof.Dr. V.Kempe, Director of the Central Institute of Cybernetics and Information Processes, and to Prof.Dr. H.Fuchs, Deputy Director of this Institute, for their efforts made in sponsoring this conference in connnection with the Society of Informatics of the GDR and the WGMA.

Our special thanks are due to Prof.Dr. D.Schmidt for his organizational work to host CAIP'87 in Wismar and to Mrs. G.Lagowitz from the Akademie-Verlag providing excellent conditions for the publication of these Proceedings.

Last, but not least, we would like to thank Mrs. E.Bahn, Mrs. H.Rustenbach, and Mr. K.-D.Müller for their untiring efforts to ensure a successful and fruitful CAIP '87.

Looking forward to CAIP '89

Azriel Rosenfeld
Leonid Yaroslavskiĭ
Wolfgang Wilhelmi

Table of Content

I. SIGNAL PROCESSING

II. ALGORITHMS

III. AI TECHNIQUES

IV. IMAGE MODELLING

V. APPLICATIONS

VI. IMPLEMENTATION

AUTHOR INDEX

DESIGN OF OPERATORS FOR IMAGE PROCESSING BASED ON THE BÉZIER-APPROXIMATION

D. Jackél, A. Koschan, H. Neumann [1)]

ABSTRACT

The function values of a two-dimensional image function can be considered as a three-dimensional surface. The often degraded image function should be preprocessed with a smoothing operator prior to a differentiation step to detect significant changes in the intensity surface. In consideration of the Bézier surface approximation technique such operators can be generated on a quadratic segment with an odd number of control points. We will call them 'Bézier' operators.

The Bernstein polynomials involved are to be understood as weighting functions of the control points. The first and second derivatives of a 'Bézier' operator can be developed easily. An efficiency evaluation of the 'Bézier' operators compared to 'Gaussian' operators suggested by Marr and Hildreth has been made. As a result, the substantially easier generation of 'Bézier'-convolution operators has to be stressed. Furthermore, empiric studies done on some test images have shown that 'Bézier' operators of smaller size are reacting less strong with regard to discretization errors than 'Gaussian' operators.

1. INTRODUCTION

For the detection of discontinuities in the image function which can be caused by different physical phenomena, Marr/Hildreth [5] suggest a set of variable sized rotationally symmetrical operators based on the Laplacian of a two-dimensional Gaussian:

$$\nabla^2 G(x,y,\sigma) = \frac{-1}{\Pi \cdot \sigma^4} \cdot \left[1 - \frac{x^2 + y^2}{2\sigma^2} \right] \cdot \exp -(x^2 + y^2)/2\sigma^2$$

The operators are scaled by the standard deviation σ. Dependent on the chosen parameters for the $\nabla^2 G$-operators the frequency spectra have bandpass characteristics with different pass ranges. The parametrization here is directly given with the number of control points responsible for the definition of the approximate Bézier surface patch. In this paper the development of discrete parametric convolution operators based on the 'Bézier'-surface approximation method is presented.

[1)] The authors are with the Institut für Technische Informatik, TU Berlin, FR3-3, Franklinstr. 28/29, 1000 Berlin 10, FRG

Furthermore, the generation of discrete convolution operators from vector functions for Bézier-surfaces is presented as well as processing results for application examples and comparisons with the results of corresponding operators derived from the two-dimensional Gaussian distribution.

2. BÉZIER-APPROXIMATION FOR DISCRETE CONVOLUTION OPERATORS

A set of picture elements can be understood as a set of control points, the value of each representing quantized intensities. The approximation of a discrete two-dimensional set of control points with a Bézier surface segment enables the smoothing of the shape of the discrete function. In the following convolution operators for image smoothing shall be developed as well as their first and second derivatives.

2.1 BÉZIER-SURFACES AND DERIVATIVES OF BERNSTEIN POLYNOMIALS

Bézier surfaces are defined as Cartesian product of two Bézier curves. The vector function based on this definition is written, according to [6]:

$$r(u,v) = \sum_{i=0}^{n} \sum_{j=0}^{m} r_{ij} \cdot B_{i,n}(u) \cdot B_{j,m}(v)$$

with

$$r_{ij} = [\ x_{ij},\ y_{ij},\ z_{ij}\ ]^T \ , \quad u,v \in [\ 0.0, \ \ldots, 1.0\ ]\ ,$$

$$B_{i,n}(u) = (n!/(i!\cdot(n-i)!))\cdot u^i \cdot (1-u)^{n-i}\ ,$$

$$B_{j,m}(v) = (m!/(j!\cdot(m-j)!))\cdot v^j \cdot (1-v)^{m-j} \quad \text{and}$$

n,m = Order of function (surface with $(n+1)x(m+1)$ control points r_{ij})

The functions $B_{i,n}(u)$ and $B_{j,m}(v)$, often referred to as 'blending functions', are Bernstein polynomials. The Cartesian components of these vector functions can be simplified because the control points are equidistant. Thus, the functions $x(u,v)$ and $y(u,v)$ are defined by the equations $x(u,v) = n\cdot u$ and $y(u,v) = m\cdot v$. With this the vector function $r(u,v)$ can be considerably reduced. We receive:

$$r(u,v) = \begin{bmatrix} n\cdot u \\ m\cdot v \\ \sum_{i=0}^{n} \sum_{j=0}^{m} z_{ij} \cdot B_{i,n}(u) \cdot B_{j,m}(v) \end{bmatrix} \tag{1}$$

With regard to the definition of rotationally symmetrical operators and the local approximation of the image intensity function by a Monge patch $r = u\cdot e_1 + v\cdot e_2 + f(u,v)\cdot e_3$ equation (1) gives

$$r(u,v) = n \begin{bmatrix} u \\ v \\ z(u,v)/n \end{bmatrix} = n \begin{bmatrix} u \\ v \\ 1/n \sum_{i=0}^{n} \sum_{j=0}^{m} z_{ij} \cdot B_{i,n}(u) \cdot B_{j,n}(v) \end{bmatrix}$$

Based on the equation (1) the first and second partial derivatives of r(u,v) can be calculated easily. Differentiating the vector function only its z-components

$$z(u,v) = \sum_{i=0}^{n} \sum_{j=0}^{m} z_{ij} \cdot B_{i,n}(u) \cdot B_{j,m}(v)$$

have to be considered. If the Bernstein polynomials are broken into their components

$$B_{i,n}(u) = C(n,i) \cdot F(u) \qquad (2)$$

with $C(n,i) = n!/(n-i)!$ and $F(u) = u^i \cdot (1-u)^{n-i}$,
the partial differentiation of first and second order of z(u,v) is reduced to the differentiation of the function F(u). We receive for the first derivative

$$F'(u) = i \cdot u^{i-1} \cdot (1-u)^{n-i} - (n-i) \cdot u^i \cdot (1-u)^{n-i-1}$$

and for the second derivative

$$F''(u) = i \cdot [(i-1) \cdot u^{i-2} \cdot (1-u)^{n-i} - (n-i) \cdot u^{i-1} \cdot (1-u)^{n-i-1}] - (n-i) \cdot [i \cdot u^{i-1} \cdot (1-u)^{n-i-1} - (n-i-1) \cdot u^i \cdot (1-u)^{n-i-2}] \qquad (3)$$

For the disassembly and the differentiation of the Bernstein polynomial as a function of the parameter v the corresponding equations result analogously.

2.2 GENERATION OF DISCRETE TWO-DIMENSIONAL CONVOLUTION KERNELS

Depending on the number of given control points a corresponding number of Bernstein polynomials are established as 'blending functions'. Inside a defined interval of the Bézier surface these define an unambiguous function value for a random point (u,v) depending on the number of control points. The influence of every control point value when calculating a specific function value at a random location (u,v) is a direct result of the functional values of the corresponding Bernstein polynomial for u or v, respectively. Thus, these have to be understood as weighting functions for the control points.

For the development of rotationally symmetrical convolution kernels the above mentioned principle can be applied. Such operator can be generated on a quadratic segment with an odd number of control points. If it is assumed that this set of picture elements for the specifically viewed local neighborhood of a point P_{ij} is the set of control points for a Bézier approximation, this point is located inside the viewed interval at the location (0.5,0.5).

For the calculation of the functional value of the Bézier surface at the location (u,v) = (0.5,0.5) the weighting of the individual control point from the Bernstein polynomials at the location u=0.5 or v=0.5, respectively, is received depending on the local neighborhood of the central picture element. Fig. 1(a) shows by way of example the shape of the Bernstein polynomials with the order n=14.

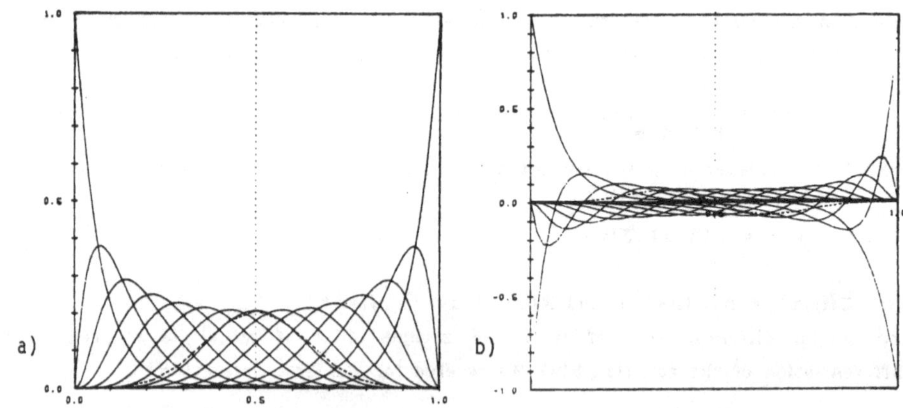

Fig. 1: Bernstein polynomials with 15 control points. The broken line in the diagrams indicates the weighting of the function values based on the values of the polynomials at u = 0.5.
(a) Original function, (b) first derivative.

For the approximation of differentiation operators from Bernstein polynomials the partial derivatives of the equation (2) can be calculated with the support of equation (3). As a result we receive the slope and/or the curvature of the polynomials for each point inside the interval. Fig. 1(b) represents the shape of the first derivative of Bernstein polynomials with $n + 1 = 15$ control points.

The convolution operators for smoothing and partial differentiation of the image function can be generated from Bernstein polynomials and their derivatives at location $(u,v) = (0.5,0.5)$. With regard to the definition of the discrete convolution (see e.g. [1]) the convolution kernels of different operators are defined as follows:

$$B(i,j) = (1/n)\cdot B_{i,n}(0.5)\cdot B_{j,n}(0.5),$$
$$B_u(i,j) = (1/n)\cdot B'_{i,n}(0.5)\cdot B_{j,n}(0.5),$$
$$B_v(i,j) = (1/n)\cdot B_{i,n}(0.5)\cdot B'_{j,n}(0.5),$$
$$B_{uu}(i,j) = (1/n^2)\cdot B''_{i,n}(0.5)\cdot B_{j,n}(0.5),$$
$$B_{vv}(i,j) = (1/n^2)\cdot B_{i,n}(0.5)\cdot B''_{j,n}(0.5), \text{ and}$$
$$B_{uv}(i,j) = (1/n^2)\cdot B'_{i,n}(0.5)\cdot B'_{j,n}(0.5).$$

For a more complete development, see [3]. Fig. 2 shows convolution operators and their partial derivatives generated by applying the 'Bézier' approximation. Fig. 3 displays by way of example the corresponding power spectra in frequency domain for B(i,j) and its Laplacian.

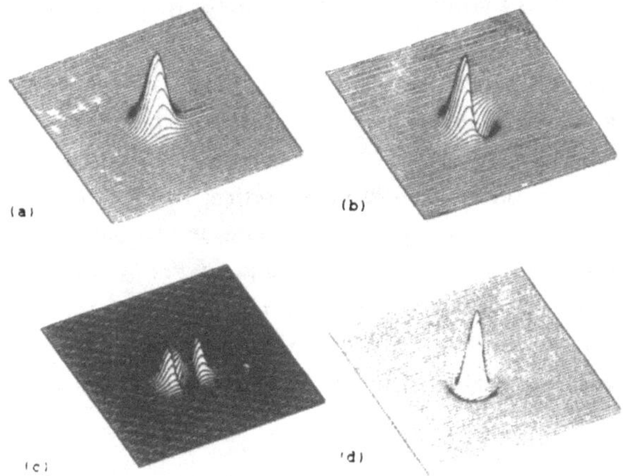

Fig. 2: Convolution operator of order n=32 generated by the method based on the Bézier approximation.
(a) Operator for noise suppression B,
(b) first partial derivative B_u,
(c) second partial derivative B_{uu},
(d) operator for generation of negative Laplacian $-\nabla^2 B = -(B_{uu}(i,j)+B_{vv}(i,j))$

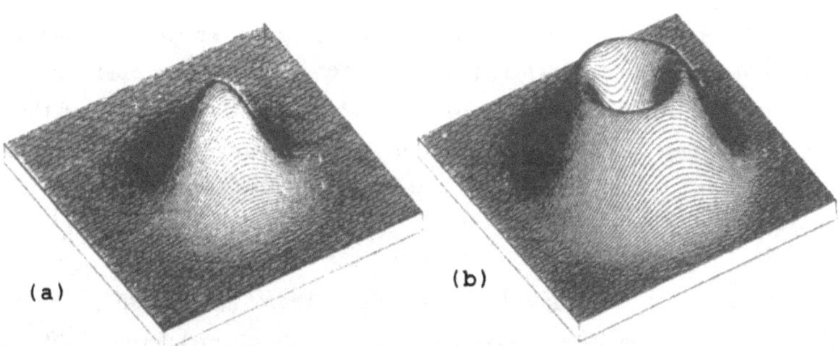

Fig. 3: Corresponding power spectra for
(a) operator for noise suppression B,
(b) negative Laplacian operator.

3. IMPLEMENTATION AND RESULTS

The 'Bézier' operators of parametric size as developed in Chapter 2 have been implemented for a floating point arithmetic. For the evaluation of the processing results from convolutions of operators with digital image matrices, operators from a two-dimensional Gaussian and their partial derivatives of equal size have been applied to the same images.

In the following, the applied operators will be presented and compared with each other. Subsequently, the image data and the processing purposes will be specified and a representation of the processing results will be given.

3.1 'BÉZIER' OPERATORS IN COMPARISON WITH GAUSSIAN OPERATORS

For the application of 'Bézier' operators for digital image matrices and the evaluation of the processing results Gaussian operators with different standard deviations were generated. $d=5.08\cdot\sigma$ as suggested by Grimson [2] served as criterion for the definition of the kernel size. For the present implementation the specific operator size was defined to be $d = \lfloor 5.08\cdot\sigma \rfloor + 4$ (i.e. functional values in the periphery are only 1% of the maximum magnitude), where the absolute value of the diameter has to be an odd number. Dependent on the parametrization of the Gaussian operators the operator sizes defined this way are serving as predetermination for the number of control points for generating corresponding 'Bézier' operators. Table 1 shows the operator parameters and their size in a survey.

Gauss		Bézier	
σ	Diameter d	Order n	Control points (n+1)
0.75	7	6	7
1.41	11	10	11
3.18	19	18	19

Table 1: Parametrization and size of applied operators

Numerical differentiation of images is an ill-posed problem which can be regularized by filtering the data prior to differentiation [8]. Small sized 'Bézier' operators for example with a diameter of 7 control points show better low pass characteristics than Gaussian operators of equal size (standard deviation σ = 0.75). Real-valued kernels and corresponding frequency spectra for equal sized (same diameter) Gaussian and 'Bézier' convolution kernels are presented in Table 2/Fig.4 and Table 3/Fig.5.

Besides the operator for the suppression of noise there have been generated specific convolution kernels for each partial derivative up to second order. With this it is possible to evaluate the magnitudes and the directions of the gradients as well as zero crossings of the second derivative apart from image smoothing.

0.00000	0.00000	0.00003	0.00009	0.00003	0.00000	0.00000
0.00000	0.00023	0.00332	0.00808	0.00332	0.00023	0.00000
0.00003	0.00332	0.04782	0.11632	0.04782	0.00332	0.00003
0.00009	0.00808	0.11632	0.28294	0.11632	0.00808	0.00009
0.00003	0.00332	0.04782	0.11632	0.04782	0.00332	0.00003
0.00000	0.00023	0.00332	0.00808	0.00332	0.00023	0.00000
0.00000	0.00000	0.00003	0.00009	0.00003	0.00000	0.00000

Table 2: Gaussian kernel 7x7 with σ = 0.75.

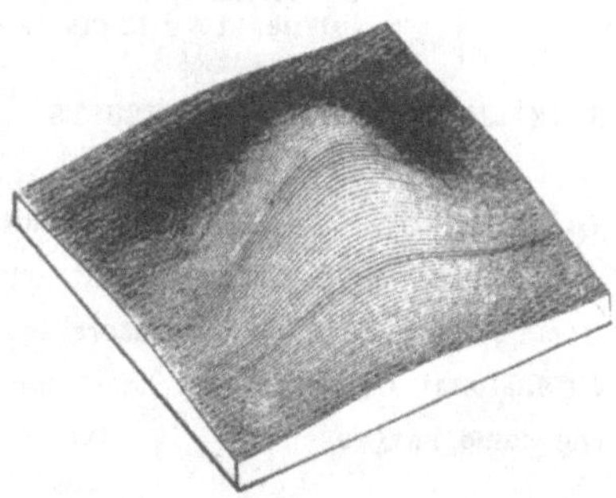

Fig.4: Plot of corresponding freqency spectra.

0.00024	0.00146	0.00366	0.00488	0.00366	0.00146	0.00024
0.00146	0.00878	0.02197	0.02929	0.02197	0.00878	0.00146
0.00366	0.02197	0.05493	0.07324	0.05493	0.02197	0.00366
0.00488	0.02929	0.07324	0.09765	0.07324	0.02929	0.00488
0.00366	0.02197	0.05493	0.07324	0.05493	0.02197	0.00366
0.00146	0.00878	0.02197	0.02929	0.02197	0.00878	0.00146
0.00024	0.00146	0.00366	0.00488	0.00366	0.00146	0.00024

Table 3: 'Bézier' kernel with 7x7 control points.

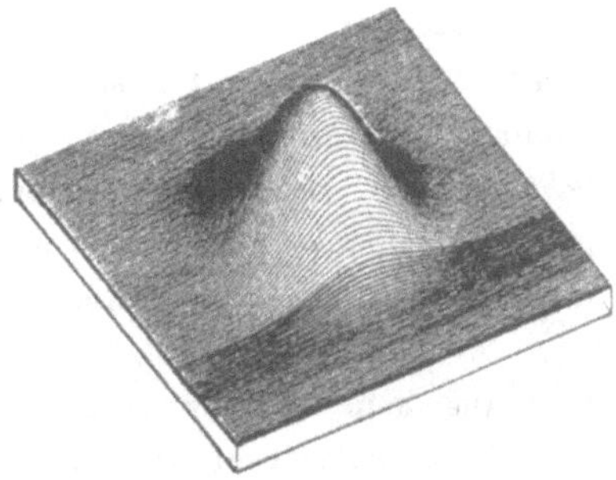

Fig.5: Plot of corresponding frequency spectra.

3.2 IMAGE DATA AND PROCESSING PURPOSES

For testing the 'Bézier' operators image matrices as described in the following chapter were applied:

CIRCRAMP_Ø is a synthetically generated image matrix where different local contrast ratios and slopes of an image function can be simulated examplarily. The function contains a ramp of constant slope, a circular region of constant intensity being located in the image center (Fig. 6(a)). Thereby the ramp is approximated by a step function of two pixels width with an increase of two intensity units each (Fig. 6(b)).

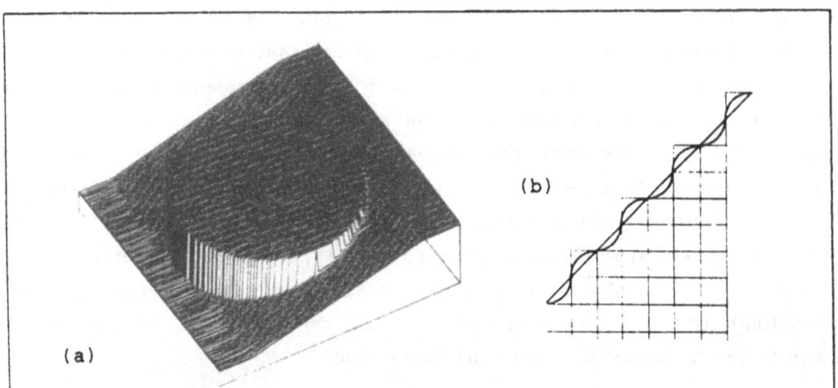

Fig. 6: Structure of the image function CIRCRAMP_Ø:
(a) surface representation of the intensity function,
(b) approximation of the ramp of constant slope.

With this image function a closed contour of constant curvature with uniformly changing local contrast ratio is simulated. The accuracy of the gradient calculation (magnitude and direction) can be tested as well as the detection of discontinuities. This test pattern can be used to evaluate the behaviour of the operator response in dependence of discretization errors and weakly contrasted images.

LSPHERE5Ø shows a synthetically generated image function of a luminance distribution for a Lambertian surface of a sphere calculated using a computer-graphics illumination model. The radius of the sphere is 5Ø pixels (equals 1Ø1 pixels in diameter). The simulated light source is in the viewer position. Viewer and light source are assumed to be at infinity, so in this case orthographic projection and parallel light rays can be assumed. For the illuminated surface of the sphere a continuous intensity surface of negative curvature is generated, bounded by a tangential cccluding contour.

With the help of this image function the susceptibility to disturbances of the convolution kernels of small size with regard to discretization errors with intensity distributions (e.g. with shaded surfaces) can be evaluated.

3.3 PROCESSING RESULTS

Fig. 7(a) and (b) show the processing results of Gaussian operators with $\sigma = 1.41$ and 'Bézier' operators with n = 1Ø on CIRCRAMP_Ø, respectively. The processing results for the visualization in Fig. 7 are strongly dependent on the selection of an interval $(-\varepsilon,+\varepsilon)$ for the definition of a real-valued zero. This arrangement is arbitrary. An 'a priori' calculation based on the image data and the convolution kernels remains an unsolved problem. The empirical determination for $\varepsilon = 1Ø^{-5}$ showed consistently good results.

The absolute values of intensity gradients vary according to the local contrast ratio continuously along the contour of the embedded region. The directions are orthogonal to the tangent of the circular contour. In regions of constant intensity the operators yield the value zero; the constant slope of the ramp causes gradients of constant norm in the direction of the steepest descend. The second derivative of the image function (Laplacian) and by this also the number of the detected zero crossings differ considerably from $\nabla^2 B * I$ to $\nabla^2 G * I$. It is recognizable that the $\nabla^2 B$ operator for the digital step function of constant slope indicates a curvature value with the absolute value of zero (linear approximation) while opposed to this the $\nabla^2 G$ operator specifically shows across a picture element alternatively positive and negative curvatures (waving approximation). This filter response therefore leads for $\nabla^2 G * I$ to the detection of zero crossings in the 'subpixel' area over the whole ramp.

Fig. 8(a) and (b) show processing results on LSPHERE5Ø for the calculation of the second derivative of curved intensity surfaces for operators of small size. For $\sigma = Ø.75$ the $\nabla^2 G$ operator shows non-significant changes in curvature (zero crossings) inside the intensity surface that has a uniform curvature. The corresponding $\nabla^2 B$-Operator with n = 6 shows only one alteration of curvature for the enclosing occluding contour.

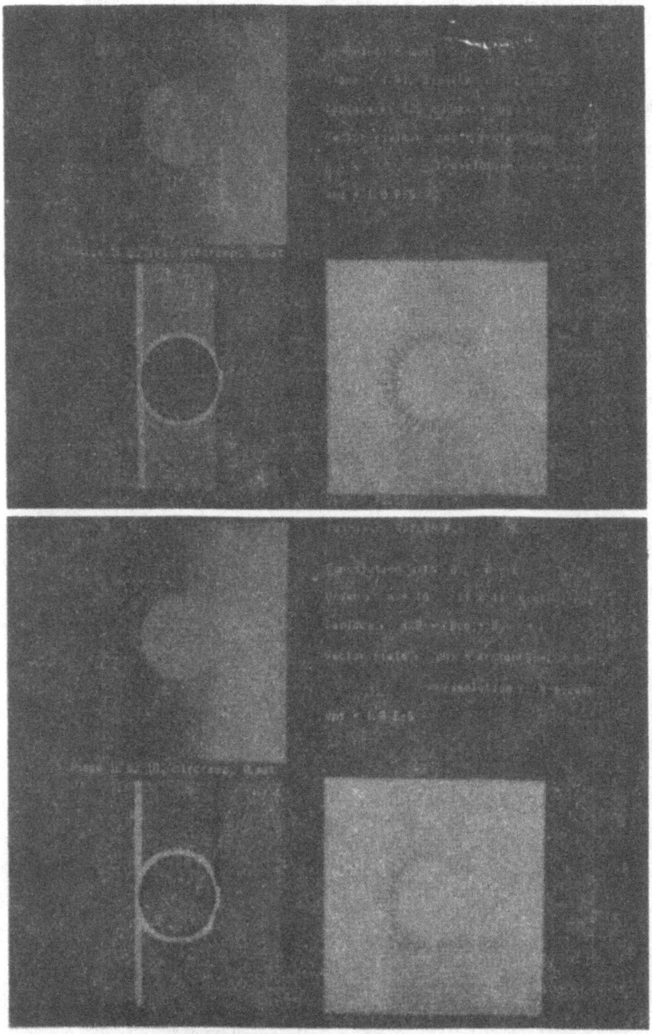

Fig. 7: Convolution of CIRCRAMP_Ø with

(a) $G*I$, $\nabla^2 G*I = (G_{xx}+G_{yy})*I$, $|\nabla G*I| = \sqrt{((G_x*I)^2+(G_y*I)^2)}$,
$\nabla_\varphi = \tan^{-1}(G_y*I/G_x*I)$ with $\sigma = 1.41$,

(b) $B*I$, $\nabla^2 B*I = (B_{uu}+B_{vv})*I$, $|\nabla B*I| = \sqrt{((B_u*I)^2+(B_v*I)^2)}$,
$\nabla_\varphi = \tan^{-1}(B_v*I/B_u*I)$ with $n = 1Ø$.

For the representation of a real-valued zero an interval $(-\varepsilon,+\varepsilon)$ with $\varepsilon = 1Ø^{-5}$ has been chosen, i.e. all values with a magnitude $|x| < 1Ø^{-5}$ have been mapped to zero.

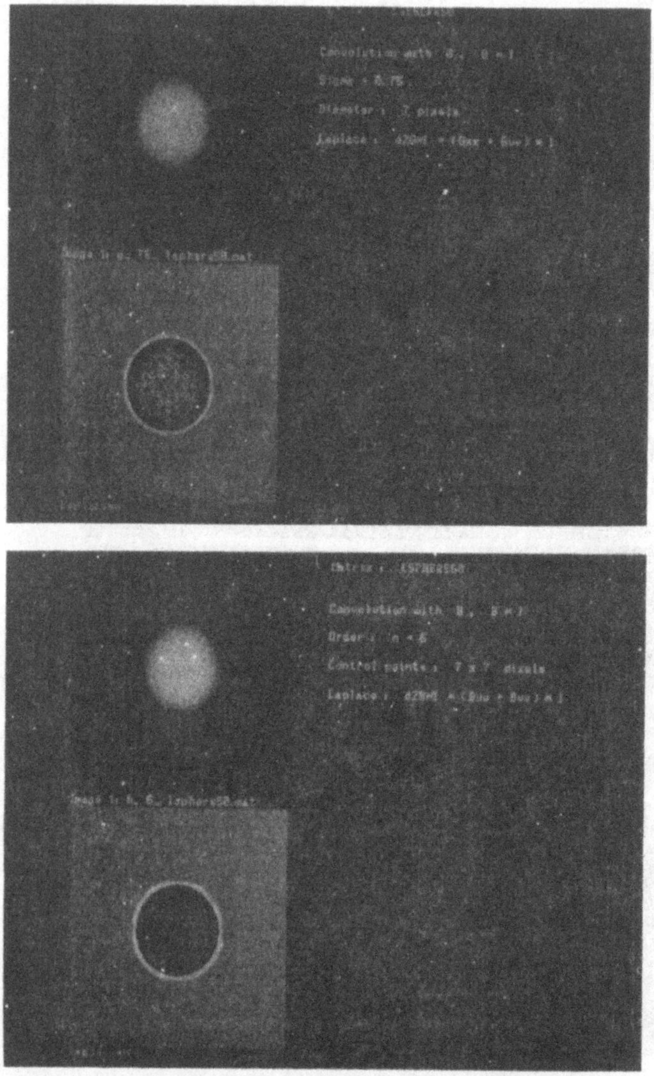

Fig. 8: Convolutions of LSPHERE5Ø with
(a) $G*I$, $\nabla^2 G=(G_{xx}+G_{yy})*I$ with σ = Ø.75
(b) $B*I$, $\nabla^2 B=(B_{uu}+B_{vv})*I$ with n = 6

4. SUMMARY

Based on a given number of control points a surface can be sufficiently approximated with the help of polynomials. We looked upon the two-dimensional set of picture elements as a set of control points representing quantized intensities. The piecewise approximation of this set of control points by a Bernstein/Bézier surface segment causes a smoothing of the discrete functional shape.

The Bernstein polynomials used for the approximation are to be understood as weighting functions of the specific discrete control points. Choosing the central element of a discrete control point interval of a Bézier surface of even order parametric operators for image processing can be generated.

The generation of 'Bézier' convolution kernels is easily practiced: Mainly, only the calculations of integer-valued factorials and of the powers of 0.5 are necessary. Thus, the generation of parametric Bézier' convolution kernels and their derivatives is numerically more stable than the generation of Gaussian kernels and their derivatives. The definition of the size of Gaussian operators depends on a randomly selected boundary for the functional value (minimum value), contrarily the size of 'Bézier' operators is uniquely defined by the number of given control points.

The processing results in Chapter 3 show that Gaussian operators of the smallest size react very strongly to discretization errors. Corresponding 'Bézier' operators, however, do not perform this way. Neurophysiological examinations of the human retina showed a sensibility profile for the receptive field which can be described in good approximation by a Gaussian distribution or a difference of two Gaussians (see f.e. Korn [4]). It has to be checked if maybe the use of Bernstein polynomials could yield to a better modeling of these profiles.

Furthermore, the less strong reaction of 'Bézier' operators with regard to discretization errors could maybe result in a more accurate operator response for local shading analysis in "lower scales" (see Pentland [7]). These aspects will be a part of our further investigations. Finally the advantages of 'Bézier' operators in comparison with Gaussian operators should be summarized:

- substantially easier generation of 'Bézier' operators,
- more numerical stability,
- less strong reaction with regard to discretization errors.

This work was carried out within SFB 203 at the Technische Universität Berlin and was funded by the Deutsche Forschungsgemeinschaft (DFG).

REFERENCES

[1] Gonzalez, R.C., P. Wintz: Digital Image Processing. Reading: Addison-Wesley 1977.

[2] Grimson, W.E.L.: From Images to Surfaces: A Computational Study of the Human Early Visual System. The MIT Press, Cambridge 1981.

[3] Jackél, D., A. Koschan, H. Neumann: Entwurf lokaler Bildvorverarbeitungsoperatoren auf der Grundlage der Bézier- Approximation. Techn. Bericht 86-11 , Institut für Techn. Informatik, TU Berlin 1986.

[4] Korn, A.: Das visuelle System als Merkmalsfilter, Fachberichte Messen, Steuern, Regeln, Band 13: Aspekte der Informationsverarbeitung, Springer Berlin 1985.

[5] Marr, D., E. Hildreth.: Theory of Edge Detection. Proc. of the Royal Society of London, Vol.207 (1980), Series B, pp.187-217.

[6] Newman, W.M., R.F. Sproull.: Principle of Interactive Computer Graphics. McGraw-Hill Tokyo, 2nd Edition 1979.

[7] Pentland, A.P.: Local Shading Analysis. Tech. Note 272, AI Center, SRI International 1982.

[8] Torre, V., T. Poggio: On Edge Detection. MIT, AI Laboratory: AI Memo 768 1984.

EDGE DETECTION ON THE BASIS OF PIECEWISE LINEAR MODELS

Herbert Jahn [1)]

1. Introduction

The detection of edges in images is one of the most important steps towards an automatical analysis of scenes. Because of the 3D - origin of 2D - images there may exist objects in the images which are not limited by closed contours (one may think about the objekt "nose" in the image of a human face). Therefore a primary forming of regions is not always the right first step and the recognition of objects primarily should be oriented at (not always closed) contours. The generation of a raw primal sketch in the sense of Marr and Hildreth /1/ therefore should be the first step in object recognition in natural scenes. This does not mean that the use of region forming methods generally fails; in simple (artificial) scenes they may work sufficiently but in complex (natural) scenes the edge detection should be primary.

Towards a simple description of scenes by edges one first has to find edge elements (edgels). In a second step these edgels must be linked to get edge fragments. In this process besides the strength of the edgels their direction is needed too. The method for the generation of edgels which is described in this paper therefore provides both the strength and direction of edge elements.

According to their different origin in the 3D - scene there exist edges of different types. The most important types are the step edge and the roof edge /2/ (Fig. 1).

Fig. 1

There exist many methods (e.g. /1/ - /4/) for the detection of edges and especially step edges. Most important are the methods basing on the gradient operator, the Hueckel operator, template matching and the zero crossings of the 2nd derivative. The foundations of these methods are certain (inflexible) models of (step-) edges and one gets bad results if the model is violated. Therefore Leclerc and Zucker /5/ propose to start from the mathematical definition of a discontinuity: "a function is discontinuous at a point if and only if its left - and right - hand limits are unequal". For instance between the points i and i+1 in an image row there is a discontinuity if the left - and right - hand limits of the grey value or one of its derivatives differ significantly.

1) AdW der DDR, Institut für Kosmosforschung
Rudower Chaussee 5, Berlin, 1199, DDR

The method proposed in this paper is basing on this philosophy. Rows and columns are handled separately and therefore it will be operated on a 1D - data set.

The edge detection on the basis of piecewise linear polynomials corrupted by white Gaussian noise is investigated in section 2. Section 3 is dedicated to the examination of the method using simulated and real image data.

2. Piecewise linear polynomials for edge detection

Be y_i (i=1...N) a given data set (e.g. image row or column), this means discrete values of a continuous (grey value) function y(x). In an interval $i \leq x < i+1$ a discontinuity may exist at any point x, but for simplicity it is looked for only in the center of the interval, that is at x = i+1/2.

It is now assumed that there is a discontinuity at x = i+1/2. In order to determine the left - and right - hand limits the function y(x) left and right of the discontinuity is approximated by a polynomial of first order

$$y = a \cdot x + b \tag{1}$$

(fig.2).

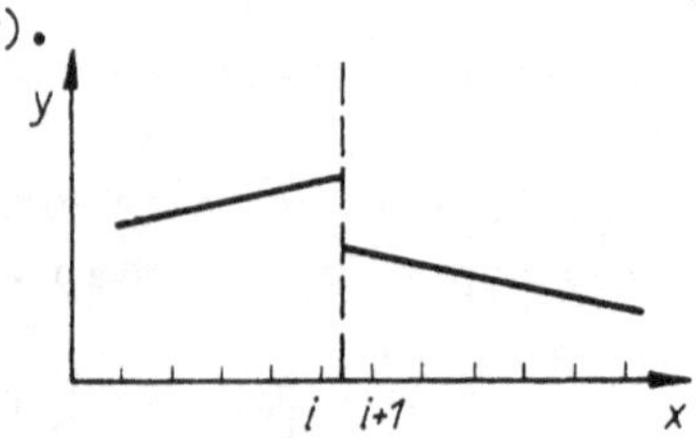

Fig. 2

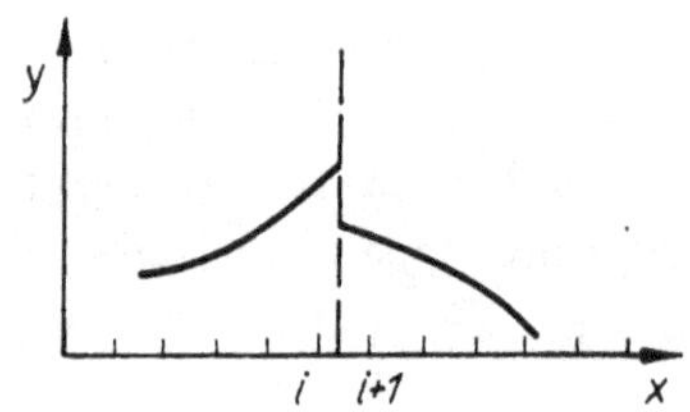

Fig. 3

This model is very flexible because it contains both step and roof edges. More general edge profils (fig. 3) can be described using piecewise polynomials of second or higher order. The piecewise linear model used here allows the determination of the left - and right - hand limits of the function y(x) and its first derivative y'(x). The values y(x) of the polynomial (1) immediately depend on y(i+1/2) and y'(i+1/2) if one makes use of the state representation (see e.g. /6/)

$$z(x) = \phi(x-x') \cdot z(x') \tag{2}$$

with the state vector

$$z(x) = \begin{pmatrix} y(x) \\ y'(x) \end{pmatrix} \tag{3}$$

and the transition matrix

$$\phi(x) = \begin{pmatrix} 1 & x \\ 0 & 1 \end{pmatrix} \tag{4}$$

To approximate real image data better the ideal grey value function (1) is corrupted by white Gaussian noise ξ_i with the expectation values

$$\langle \xi_i \rangle = 0 \quad \text{and} \quad \langle \xi_i \cdot \xi_j \rangle = \sigma^2 \cdot \delta_{ij} . \tag{5}$$

Then the data values y_j are expressed by

$$y_j = H^T \cdot z(j) + \xi_j \tag{6}$$

with $H^T = (1,0)$.

Using equation (2) y_j can be written in dependence of z(i+1/2):

$$y_j = H^T \cdot \phi(j-i-1/2) \cdot z(i+1/2) + \xi_j . \tag{7}$$

This model is the basis for the estimation of the left - and right - hand state vectors $z^L(i+1/2)$ and $z^R(i+1/2)$.

For the estimation of the left - hand state vector $z^L(i+1/2)$, which is now considered, the left - hand data values $y_i, y_{i-1}, \dots, y_{i-n_L+1}$ are used. The number n_L of the used data values at the moment is assumed to be a fixed value. The Maximum - Likelihood - estimation of $z^L(i+1/2)$ leads to the minimization of the function

$$\begin{aligned} J^L_{n_L}(z^L(i+1/2)) &= \sum_{j=i-n_L+1}^{i} \left[y_j - H^T \phi(j-i-1/2) \cdot z^L(i+1/2)\right]^2 \\ &= \sum_{k=1}^{n_L} \left[y_{i+1-k} - H^T \phi(1/2-k) \cdot z^L(i+1/2)\right]^2 \end{aligned} \tag{8}$$

with the solution

$$\hat{z}^L_{n_L}(i+1/2) = (M^L_{n_L})^{-1} \cdot f^L_{n_L}(i) . \tag{9}$$

Here the 2 - vector $f^L_{n_L}(i)$ is given by

$$f^L_{n_L}(i) = \sum_{k=1}^{n_L} \phi^T(1/2-k) \cdot H \cdot y_{i+1-k} \tag{10}$$

and the 2 x 2 - matrix $M^L_{n_L}$ by

$$M^L_{n_L} = \sum_{k=1}^{n_L} \phi^T(1/2-k) \cdot H \cdot H^T \cdot \phi(1/2-k) . \tag{11}$$

Using (10) the estimation values (9) can be written by

$$\hat{z}^L_{n_L}(i+1/2) = \sum_{k=1}^{n_L} a^L_{n_L,k} \cdot y_{i+1-k} \tag{12}$$

with the coefficients (2 - vector)

$$a^L_{n_L,k} = (M^L_{n_L})^{-1} \cdot \phi^T(1/2-k) H . \tag{13}$$

(12) is a finite impulse response (FIR) filter for the calculation of the left - hand limits $\hat{y}^L(i+1/2)$ and $\hat{y}'^L(i+1/2)$ which is easy to implement on a digital computer.

The estimation quality of the νth state component $z^L_\nu(i+1/2)$ (ν = 1,2) is given by the mean quadratic estimation error

$$Q^L_{n_L,\nu}(i+1/2) = \left\langle \left[\hat{z}^L_{n_L,\nu}(i+1/2) - z_\nu(i+1/2)\right]^2 \right\rangle . \tag{14}$$

From (7) and (12) follows the equation

$$\hat{z}^L_{n_L,\nu}(i+1/2) = z_\nu(i+1/2) + \sum_{\mu=1}^{2} (M^L_{n_L})^{-1}_{\nu,\mu} \sum_{\kappa=1}^{n_L} \sum_{\lambda=1}^{2} \phi^T_{\mu\lambda}(1/2-k)\cdot H_\lambda \cdot \xi_{i+1-k}$$

and , using (5),

$$Q^L_{n_L,\nu}(i+1/2) = \sigma^2\cdot(M^L_{n_L})^{-1}_{\nu,\nu} \qquad (\nu = 1,2) \,. \tag{15}$$

Because the noise variance σ^2 is not known the estimation value

$$(\hat{\sigma}^L_{n_L})^2 = \frac{1}{n_L}\sum_{\kappa=1}^{n_L} \left[y_{i+1-k} - H^T\cdot\phi^T(1/2-k)\cdot\hat{z}^L_{n_L}(i+1/2)\right]^2$$

$$= \frac{1}{n_L}\, J^L_{n_L}(\hat{z}^L_{n_L}(i+1/2)) \tag{16}$$

must be used. Then the left - hand estimation error of the νth state component is given by

$$\varrho^L_{n_L,\nu}(i+1/2) = \left[\hat{Q}^L_{n_L,\nu}(i+1/2)\right]^{1/2} = \left[(M^L_{n_L})^{-1}_{\nu,\nu}\cdot\frac{1}{n_L}\cdot J^L_{n_L}(\hat{z}^L_{n_L}(i+1/2))\right]^{1/2}. \tag{17}$$

$$(\nu = 1,2)$$

The right - hand estimation values $\hat{z}^R_{n_R}(i+1/2)$, $\varrho^R_{n_R,\nu}(i+1/2)$ one obtains in an analoguous manner processing the data set y_i (i=1...N) from right to left, which means using the inverted data set $\check{y}_j = y_{N-j+1}$ (j=1...N).

Now it can be decided whether there is a discontinuity or not. Writing $\hat{y}^L = \hat{z}^L_{n_L,1}(i+1/2)$, $\varrho^L = \varrho^L_{n_L,1}(i+1/2)$, $\hat{y}^R = \hat{z}^R_{n_R,1}(i+1/2)$, $\varrho^R = \varrho^R_{n_R,1}(i+1/2)$ and assuming $\hat{y}^L > \hat{y}^R$ it follows from fig. 4 that there is a discontinuity if the inequality

$$\hat{y}^L - K\cdot\varrho^L > \hat{y}^R + K\cdot\varrho^R$$

or $\hat{y}^L - \hat{y}^R > K\cdot(\varrho^L + \varrho^R)$ is fulfilled. Here K is a certain parameter.

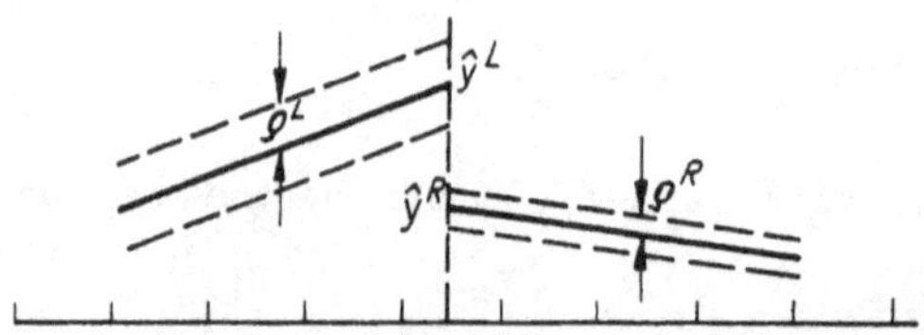

Fig. 4

In the general case the inequality

$$|\hat{y}^R - \hat{y}^L| > K\cdot(\varrho^R + \varrho^L)$$

or (for both state components)

$$|\hat{z}^R_{n_R,\nu} - \hat{z}^L_{n_L,\nu}| > K_\nu\cdot\left[\varrho^R_{n_R,\nu} + \varrho^L_{n_L,\nu}\right] \qquad (\nu = 1,2) \tag{18}$$

must be fulfilled. One obtains discontinuities of 0th order if (18) is valid for $\nu = 1$ and discontinuities of 1st order for $\nu = 2$.
If there is a discontinuity of 0th order at x=i+1/2 but not one of 1st order, then it corresponds to a step edge (fig. 5a). Reversely, a roof

edge corresponds to a discontinuity of 1st order (fig. 5b). If there are discontinuities of 0th and 1st order then one has a combined edge type (fig. 5c).

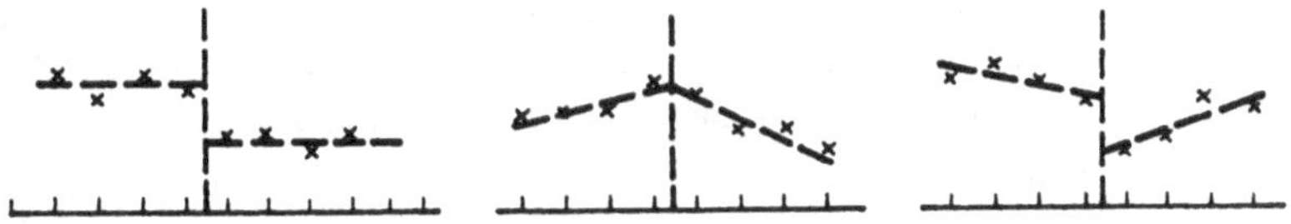

a) b) c) Fig. 5

Now it must be discussed how to choose the numbers n_L and n_R which define the extent of the left - hand and right - hand estimation interval respectively. Obviously n ($n = n_L, n_R$) must be grater than or equal two in order to make possible the estimation of two unknowns z_1 and z_2. A reduction of noise results for $n \geq 3$ only. But an unlimited increase of n in general is not possible because then the polynomial model becomes invalid. According to /7/ one gets an optimal value $n_{L,opt}$ of n_L (and analogously of n_R) if one looks for the minimum of the estimation error $\varrho^L_{n_L,\nu}(i+1/2)$ with respect to n_L (for $n_L \geq n_{L,min} \geq 3$). In order to do that the quantities $\varrho^L_{n_L}$ must be calculated for all possible values $n_L \geq n_{L,min}$. This can be done recursively if one makes use of the equations

$$f^L_{n_L+1}(i) = f^L_{n_L}(i) + \phi^T(-n_L-1/2)\cdot H\cdot y_{i-n_L} ,$$

$$M^L_{n_L+1} = M^L_{n_L} + \phi^T(-n_L-1/2)\cdot H\cdot H^T\cdot \phi(-n_L-1/2) ,$$

which immediately follow from (10) and (11) (see /7/).
But much computer time is needed if one uses this method, and therefore it is only justified if very weak edges have to be detected. In most cases it is better to choose fixed values n_L, n_R in dependence of the class of pictures which are to be processed.

Now the advantages and disadvantages of the method have to be discussed. First of all, as already has been said, the method enables one to detect edges of a general type with step and roof edges as special cases. In large regions with smoothly increasing or decreasing grey values no edges will be found. In this respect the method is superior to the well - known gradient methods. The possibility of detection of an edge according to (18) depends on the dissimilarities in the vicinity of the edge. The less the deviations of the grey values from linear polynomials are, that is the less the quantities $\varrho^R_{n_R,\nu}$ and $\varrho^L_{n_L,\nu}$ are, the less may be the difference $|\hat{z}^R_{n_R,\nu} - \hat{z}^L_{n_L,\nu}|$ which represents the strength of the edgel. So very weak edges may be detected if the data values left - hand and right - hand of the discontinuity fit the polynomials exactly. To avoid the detection of too weak edges one may change the decision criterion (18) slightly to get

$$|\hat{z}^R_{n_R,\nu} - \hat{z}^L_{n_L,\nu}| > K_\nu \cdot \mathrm{MAX}\{\Delta_\nu ; \varrho^R_{n_R,\nu} + \varrho^L_{n_L,\nu}\} . \qquad (19)$$

Here the quantities Δ_ν (ν = 1,2) are certain thresholds which define lower limits for the detection of discontinuities of 0th and 1st order respectively. According to (19) one can define a quntity

$$S_\nu = \frac{\hat{z}^R_{n_R} - \hat{z}^L_{n_L}}{\mathrm{MAX}\{\Delta_\nu ; \varrho^R_{n_R,\nu} + \varrho^L_{n_L,\nu}\}} , \qquad (20)$$

which may be associated to an edgel as a feature with $|S_\nu|$ being the strength of the edgel of order ν-1. If one has found a discontinuity in row i at position j with a strength $|S^{row}_\nu|$ and another one in column j at position i with a strength $|S^{col}_\nu|$ then it is possible to defind an edgel at (i,j) which is characterized by the feature vector $(S^{row}_\nu , S^{col}_\nu)$ giving the strength and direction of the edgel.
Another good property of the method follows from the possibility of an optimal choice of the numbers n_R and n_L. At a given noise strength the capability of detection of a discontinuity increases with increasing n_R and n_L as long as the used polynomial model is not violated. This is a property which the visual system of humans has too: The perception of an edge increases with the extends of the adjacent homogeneous regions.
Another advantage is the possibility of a straightforward extension of the method to the detection of discontinuities of higher order using polynomials of order ν-1 ($\nu \geq 3$).
The main deficiency of the method follows from the fact that the used model does not represent the reality sufficiently good. Sharp discontinuities usually do not occur in real images because the modulation transfer function of the receiving camera provides a blur. Therefore in most cases "soft" edges will be observed, and in the next step the convolution of the polynomial pieces with the line spread function of the used camera must be provided.
A further deficiency of the method results from the used one - dimensional half - neighborhoods. If n_L and / or n_R are small then the reduction of noise may be not sufficient resulting in false edgels. Therefore a generalization of the method to two - dimensional half - neighborhoods would be desirable.

3. Results

First of all the method has been tested using simulated data. Fig. 6 shows a step - like grey value profile corrupted by uncorrelated Gaussian noise with different dispersions (σ = 0.1 for $i \leq 40$, σ = 0.2 for $i \geq 41$).

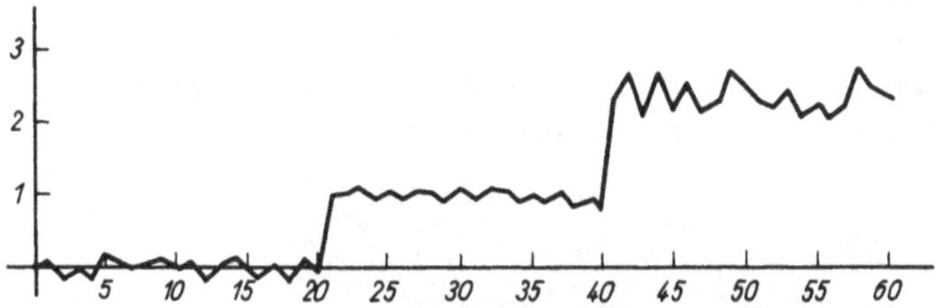

Fig. 6

Various combinations of the thresholds K_ν and Δ_ν has been checked. In the case of fig. 6 (and in other cases too) one obtains good results for $K_\nu = 3...5$ ($\nu = 1,2$), $\Delta_1 = 0...0.3$ and $\Delta_2 = 0...0.03$. Choosing $n_L = n_R = 5$ and $K_1 = 3$, $K_2 = 4$, $\Delta_1 = 0.3$, $\Delta_2 = 0.03$ one finds discontinuities of 0th order between 20 and 21, 40 and 41 and discontinuities of 1st order between 17 and 18, 18 and 19, 22 and 23. The latter disappear if one makes use of the algorithm with optimal choice of n_L and n_R ($n_L, n_R \geq 5$). Then discontinuities of 0th order between 19 and 20, 20 and 21, 38 and 39, 40 and 41, 41 and 42 will be found (with the greatest strength between 20 and 21, 40 and 41).

Fig. 7 shows a discontinuity of 1st order, that is $y_i = \sigma \cdot \xi_i$ (i = 1..40), $y_i = -2 + 0.05 \cdot i + \sigma \cdot \xi_i$ (i = 41...80), $\sigma = 0.1$. Using $n_L = n_R = 5$ and the same parameters K_ν, Δ_ν as before one does not find a discontinuity. Covering the ranges $i < 35$ and $i > 45$ with a sheet of paper each the visual system of humans does not find the discontinuity too. But the algorithm with the optimal choice of n_L and n_R ($n_L, n_R \geq 5$) finds discontinuities in the vicinity of i = 40, and the one with the greatest strength is situated between 39 and 40. At this discontinuity the optimal values of n_L and n_R are 41 and 39 respectively. This means that all data have been used for the estimation.

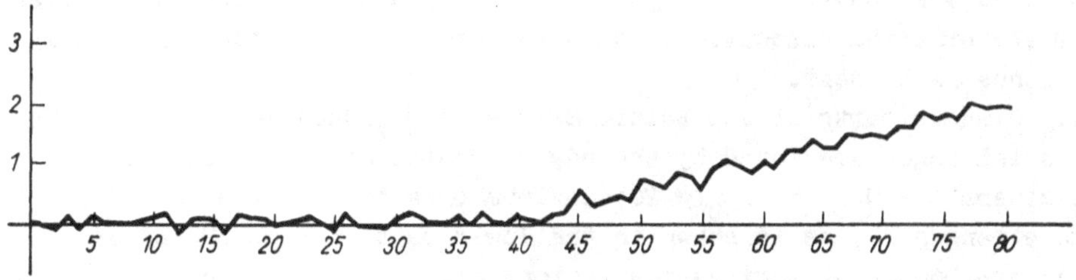

Fig. 7

In the upper part of fig. 8 a simulated grey value step (on the left - grey value 190, on the right - 210) corrupted by white Gaussian noise with the dispersion $\sigma = 5$ is shown. The lower part of fig. 8 shows the overlay of the result of the application of the 0th order edge operator with the parameters $K_1 = 3, \Delta_1 = 0.3$. The optimal values of the numbers n_L and n_R automatically have been chosen within the interval [5,20].

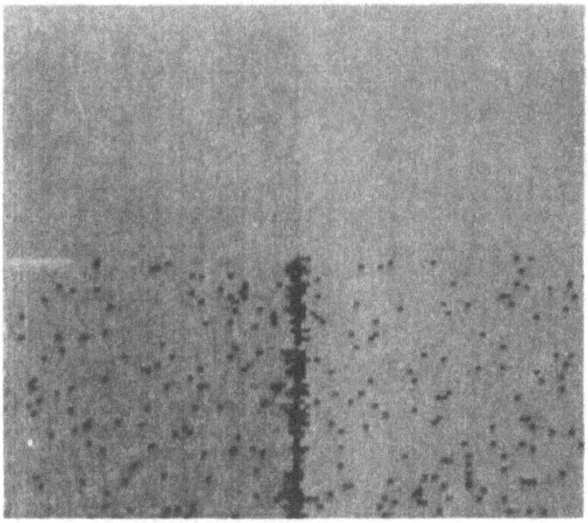

Fig. 8

The upper part of fig. 9 represents another simulated scene. In the columns 1 - 44 and 85 - 128 there is a constant intensity with mean grey value 100. From column 45 to column 64 the grey value linearly increases up to the top level 200. After that it linearly decreases, reaching the value 100 at column 84. This spike as well as the other parts of the image again are corrupted by white Gaussian noise with $\sigma = 8$. It is difficult to recognize the discontinuity of 1st order at column 64 in the photographic picture which means that it is not a very strong one. The lower part of the picture displays the strength of the found discontinuities of 1st order (maximum brightness corresponds to a strength of 15, $\Delta_2 = 0.02$, $5 \leq n_{opt} \leq 20$). In almost all image rows the existing discontinuities have been found but the strong noise influence is evident.

Fig. 10 shows parts of the Baltic Sea islands Hiddensee and Rugia. The essential edges are found by the edge operator of 0th order with the parameters $\Delta_1 = 0.3$, $5 \leq n_{opt} \leq 20$ (maximum brightness corresponds to edge strength 8), as is shown in the lower part of fig. 10. Edges which are bounds of very narrow regions are not detected satisfactorily because $n_{min} = 5$ was chosen. With respect to this better results will be obtained if one chooses $n_{min} = 3$. But then many additional noise edgels arise.

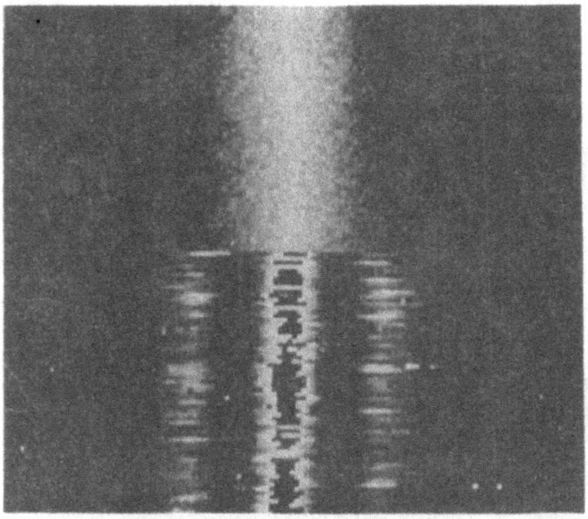

Fig. 9

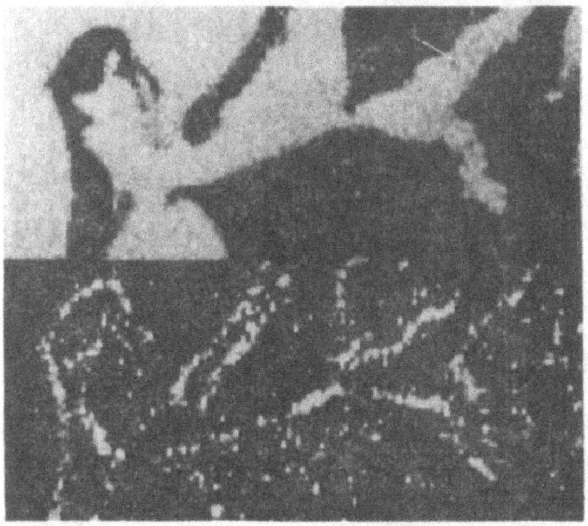

Fig. 10

The aerial photograph of an agricultural field (fig. 11) shows stripe - like structures which very much resemble the simulated spike of fig. 9. Applying the 0th order edge operator one does not obtain useful results but the 1st order edge strength displayed in the lower part of the picture satisfactorily reproduce the structure ($\Delta_2 = 0.02$, $5 \le n_{opt} \le 10$). It follows that in natural scenes roof edges may be essential which means that the use of the usual edge operators for the detection of step edges is not sufficient.

The last picture (fig. 12) has been generated by computer graphics. It contains regions with gradually changing grey values as well as edges of different type and strength. The lower part of the picture shows discontinuities of 0th order. Maximum brightness corresponds to an edge strength of 10 ($\Delta_1 = 0.3$, $5 \le n_{opt} \le 20$). Weak edges which cannot be reproduced by the photographic picture (e.g. in the right upper part) precisely have been found and the bright central point within the circle on the left of the picture too.

Fig. 11

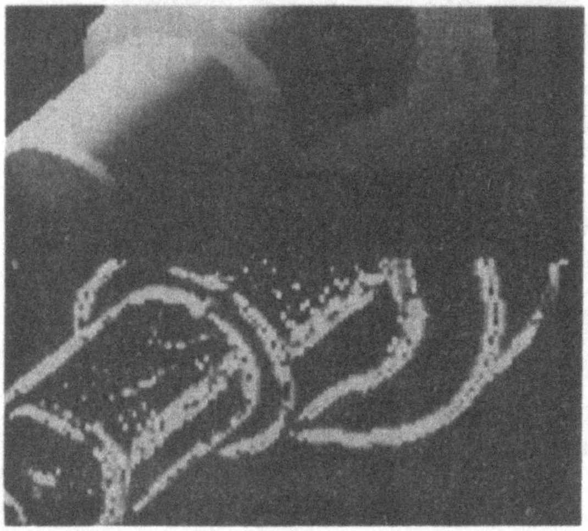

Fig. 12

REFERENCES

/1/ D. Marr, E. Hildreth; Theory of edge detection; Proc. R. Soc. Lond. B 207, 1980,p. 187 - 217

/2/ R. M. Haralick; Digital step edges from zero crossing of second directional derivatives; IEEE Trans., PAMI-6, 1984, p. 58 - 68

/3/ G. B. Shaw; Local and regional edge detectors: some comparisons; Computer Graphics and Image Processing 9, 1979, p. 135 - 149

/4/ R. Nevatia; Machine Perception; Prentice - Hall, Inc., Englewood Cliffs, N. J., 1982

/5/ Y. Leclerc, S. W. Zucker; The local structure of image discontinuities in one dimension; Proc. of the 7th Int. Conf. on Pattern Recogn., Montreal, 1984

/6/ L. A. Zadeh, C. A. Desoer; Linear System Theory; Mc Graw - Hill, New York, 1963

/7/ H. Jahn; Zustandsschätzung mit adaptivem Gedächtnis; MESSEN STEUERN REGELN 24, 1981, S. 146 - 149

RECURSIVE ALGORITHMS FOR LOCAL ADAPTIVE LINEAR FILTRATION

R. Vitkus, L. Yaroslavsky*

Introduction

The class of adaptive linear filters for picture processing as described in [1,2] features frequency response adjustable to the observed spectrum of the picture to be processed. These filters can be adaptive locally or globally. With global adaptation, the filter parameters are adjusted to the spectrum estimate of all the picture which is processed as a whole. With local adaptation, the filter parameters are adjusted to the spectra of individual image fragments, and the image is processed on the fragment-by-fragment basis. Finally, with sliding processing where the passage from one fragment to another is done by shifting the fragment center by one sample, at each step only the central sample is estimated and the adjustment is performed over all the fragment. This implies that in order to perform filtration one must estimate the spectrum of each picture fragment, generate the filter frequency response in terms of the spectrum, multiply it by the fragment spectrum and perform the inverse spectral transform in a single point, the central fragment sample.

The technique of adaptive filter parameter adjustment is independent of the chosen orthogonal basis [1, 2] with respect to which picture fragment spectrum is estimated. Filtration effectiveness, therefore, is defined by the effectiveness of the algorithms for computation through picture fragment spectra. The discrete Fourier transform (DFT) is one of the most common orthogonal transforms used in picture processing. The two-dimensional DFT is known (see, for example, [2]) to be computable recursively which dramatically reduces the amount of computations in sliding processing. However, the DFT-based estimation of Fourier spectra has disadvantages due to the pronounced influence of the boundary effects on the spectrum. Special spectral windows provide better spectral estimation. The discrete cosine transform (DCT) is an alternative to spectral windows. It makes better use of signal samples for spectrum estimation. Recursive spectral analysis algorithms based on DFT with spectral windows or on DCT have never been described in the literature. This paper is devoted to the study of the class of linear orthogonal two-dimensional transforms for which recursive spectral ana-

* The authors are with the Institute of Information Transmission Problems, at the USSR Academy of Sciences, 19, Ermolovoy st. Moscow, 101447, USSR

lysis algorithms can be constructed and to their description for DFT with cosine spectral window and for DCT.

The class of base functions and recursive algorithms of local spectral analysis.

We confine our consideration to two-dimensional linear transforms that are performed over rectangular picture fragments and are separable. All the widely used picture-independent two-dimensional orthogonal transforms meet this constraint by construction. Starting with the property of separability, the two-dimensional recursive algorithms may be constructed by means of one-dimensional transforms along picture rows and columns. Therefore, let us consider at first the algorithms of one-dimensional recursive transforms.

Let $\{a(n)\}$ be sequence of, generally, complex numbers, and $\{\varphi_r(k)\}_{r=0}^{M-1}$ be a set of base functions of a transform that are orthogonal over an interval of length M. The components of spectal sample vector $\alpha(n)=\{\alpha_r(n),\ r=0,1,\ldots,M-1\}$, as computed through the fragment $\{a(n-M+1),a(n-M+2),\ldots,a(n)\}$ are defined by

$$\alpha_r(n)=\sum_{k=0}^{M-1} a(n-M+k)\,\varphi_r(k)\ ,\quad r=0,1,\ldots,M-1 \tag{1}$$

Write this expression as

$$\alpha_r(n)=\sum_{k=0}^{M-1} a(n-k)\,h_r(k)\ ,\quad r=0,1,\ldots,M-1 \tag{2}$$

where $h_r(k)=\varphi_r(M-1-k)$. It describes M linear filters with finite pulse response that compute spectrum samples when the processed fragment slides along the sequence.

Let us determine the form of filter pulse responses allowing recursive determination of the coefficients $\alpha_r(n)$ with respect to n. At the sample-by-sample shifting of the processed fragment, one end sample leaves it and another appears on the opposite side. Therefore, the requirement that the spectral samples be computed as a linear combination of (a) input and output of the initial sequence at shifting the fragment and (b) the value of the same spectral samples in the previous position of the processed fragment

$$\alpha_r(n)=c_r\alpha_{r-1}(n-1)+b_r a(n)+d_r\, a(n-M)\ ,\quad r=0,1,\ldots,M-1\ , \tag{3}$$

where $\{c_r\}$, $\{b_r\}$ and $\{d_r\}$ are arbitrary numbers, is one of the simplest and natural recursiveness conditions.

According to (3), the spectral samples $\{\alpha_r(n)\}$ are computed independently of each other, and at most three multiplications and two additions of, generally, complex numbers are required for one sample. The following expressions may be written for the pulse responses of linear systems described by the difference equations (3):

$$h_r(k) = \begin{cases} b_r c_r^k , & k = 0,1,\dots, M-1 \; ; \\ c_r^{k-M}(b_r c_r^M + d_r), & k \geqslant M \; . \end{cases}$$

Since we are interested in the final pulse responses of length M , the condition

$$b_r c_r^M = - d_r$$

must be obeyed. Finally, obtain the following relations for $\{h_r\}$ and the coefficients of (3):

$$h_r(k) = b_r c_r^k \; ; \; d_r = -b_r c_r^M \; ; \; k = 0,1,\dots, M-1 \; ,$$

where b_r and c_r are arbitrary numbers. Hence, the following constraints on $\{\varphi_r(k)\}$ result:

$$\begin{aligned} \varphi_r(k) &= \varphi_r(0)\, c_r^{-k} \; ; \; k = 0,1,\dots, M-1 \; ; \\ d_r &= -\varphi_r(0)\, c_r \; ; \\ b_r &= \varphi_r(0)\, c_r^{1-M} \; , \end{aligned} \tag{4}$$

where $\{\varphi_r(0)\}$ and $\{c_r\}$ are arbitrary numbers. As one may see from (4), the recursiveness condition (3) requires that each base function be a segment of geometric progression. Obtain from the abovesaid and from the orthogonality condition of $\{\varphi_r(k)\}$ that $\varphi_r(k)$:

$$\varphi_r(k) = u_r \exp[i2\pi(r+R)(k+K_r)/M] \; , \quad k, r = 0,1,\dots, M-1 \; ,$$

where $\{u_r\}$, $\{K_r\}$ and R are arbitrary real numbers. If the system $\{\varphi_r(k)\}$ is orthonormal, $u_r = \pm 1/\sqrt{M}$. The choice of the parameters $\{K_r\}$, $\{u_r\}$ and R defines a particular base. If for all r one substitutes $K_r = K$ and $u_r = 1/\sqrt{M}$, obtain the known base of shifted DFT (SDFT) with parameters K and R defining the shift in the space and in the spectral domain, respectively, ([2]). In this case, the parameters $\{\varphi_r(0)\}$ and $\{c_r\}$ from (4) are

$$\begin{aligned} \varphi_r(0) &= (1/\sqrt{M}) \exp[i2\pi(r+R)K/M] \; , \\ c_r &= \exp[-i2\pi(r+R)] \; . \end{aligned} \tag{5}$$

By substituting them into (4) obtain after appropriate transformations the following recursive algo ithm for computation of SDFT spectrum $\{\alpha_r(n)\}$:

$$\alpha_r(n) = (1/\sqrt{M})\{\alpha_r(n-1) + \exp[i2\pi(r+R)K/M] \; \underset{\smile}{x}$$

$$\underset{\sim}{x} \; (\exp(i2\pi R)a(n)-a(n-M))\} \exp[-i2\pi(r+R)/M] \tag{6}$$

Notably, the amount of arithmetic operations for computation of $\alpha_r(n)$ depends on the particular choice of K and R .

As it may be seen from (4), only a narrow class of base functions of orthogonal transforms satisfies the recursiveness condition (3). To extend the class, one has to relax the recursiveness condition. A possible way to generalization is defined as follows:

$$\begin{aligned} \alpha_r^{(p)}(n) &= c_r^{(p)} \alpha_r^{(p)}(n-1) + b_r^{(p)} a(n) + d_r^{(p)} a(n-M) \; , \; p=1,2,\dots P \; , \\ \alpha_r(n) &= \sum_{p=1}^{P} \alpha_r^{(p)} \quad ; \quad r=0,1,\dots,M-1 \end{aligned} \tag{7}$$

This condition simply states that each spectral coefficient must be computed as a sum of two variables each of which is recursively computed through (3). Obviously, in practice P should not be too large. Below we discuss the case of P =2.

From the comparison of (7) and (4) one may see that for P =2 the base functions of the transforms satisfying (7) are the sums of segments of two geometric progressions. One can readily see that among these transforms are the well known discrete cosine and sine ones ([2]). In order to compute the spectra of these transforms there is, of course, no need, to perform all the operations defined by (7). For example, the DCT of real sequences is computable through

$$\alpha_r(n) = Re\left\{\sum_{k=0}^{M-1} a(n-M+1+k)\exp[i2\pi(2k+1)r/2M\right\} \tag{8}$$

which, in its turn, may be computed recursively through (3).

A still more effective algorithm can be obtained if one regards (7) as describing the linear system computing spectral samples $\{\alpha_r\}$ by a system of difference equations in the state space where $\{\alpha_r^{(1)}, \alpha_r^{(2)}, \dots, \alpha_r^{(p)}\}$ are state variables. By excluding the state variables one can obtain a higher-order difference equation relating system's input and output. In the case of DCT, obtain an output second-order difference equation where all the coefficients are real numbers. In spite of higher equation order, the amount of arithmetic operations may be lower than in the case of (3) where the equation coefficients are complex-valued. Algorithms of this kind exist also for the discrete sine transform and other transforms obtained from SDFT for symmetrical fragments ([2]).

Consider now algorithms for computation of two-dimensional spectra for MxN-dimensional fragmentnts where M and N are numbers of columns and rows, respectively. There are two approaches to the computation of two-dimensional separable transforms - by using the above one-dimensional algorithms along one coordinate or both coordinates.

The first approach is recommendable for computers with small main memory. The recursive algorithm is used only for transform computations along fragment rows when sliding along picture rows. Understandably, to compute the two-dimensional spectrum one needs at each shift the spectra of the new column and of some preceeding columns that may be stored in the buffer memory. The column-wise transforms are performed through fast orthogonal transforms or direct computations. The merit of this technique is comparatively small main memory area required for partial results because the number of fragment column spectra to be stored is of the order of fragment row length. Among the demerits of this technique one may cite the ineffectiveness of column-wise transforms if the columns are large enough as well as the constraints on column height imposed by some fast transformation algorithms.

If the computer memory can accomodate the transforms of all columns of the fragments along a picture row, the column spectra may be determined also through the one-dimensional recursive algorithms.

Spectral window-based recursive algorithms.

The spatial filtration by means of the discrete Fourier transform in the frequency domain leads to boundary effects due to DFT cyclicity ([2]). One of the tools for elimination of these effects is the so called apodization, i.e. element-wise multiplication of the signal by the window apodizing function that is equal to one in the central part of the signal sequence and vanishes towards its edges. With the recursive algorithms under consideration, the form of the function, obviously, should be chosen so as not to handicap recursive determination of the spectra of the neighbouring fragments.

One may easily see that the two-dimensional apodizing window functions that do not violate the recursiveness condition (3) are as follows:

$$\rho(k,l) = \begin{cases} c\, x^k y^l, & m_1 \leq k \leq m_2 \; ; \; n_1 \leq l \leq n_2 \; ; \\ 0 \;, & 0 \leq k < m_1 \; ; m_2 < k < M ; 0 \leq l < n_1 ; n_2 < l < N, \end{cases}$$

where c , x and y are arbitrary complex numbers. Linear combinations of these functions, obviously, also do not violate (7). These functions include complex exponents and their linear combinations, i.e. functions described by products of sines and cosines. Consider a recursive algorithm for determination of SDFT for apodization by cosine window. The two-dimensional cosine window is defined as follows:

$$\rho(k,l) = \cos[2\pi(k-T)d/M] \cos[2\pi(l-U)g/N] \quad ,$$

$$k = 0, 1, \dots, M-1 \; ; \; l = 0, 1, \dots, N-1$$

The position of window maximum in picture fragment and the steepness of weighing window sample decrease towards the edges may be varied by an appropriate choice of T , u , d and g .

Consider a particular SDFT with cosine window apodization that allows for the fact that the picture elements are real-valued. In the one-dimensional case of transforming fragment columns or rows it is described by

$$\alpha_{r}(n) = \sum_{k=0}^{M-1} a(n-M+1+k) \cos[2\pi(k-K)d/M] \exp[i2\pi(k-K)r/M] \; .$$

In this case the most effective algorithm results from the substitution of cosine exponent whose real part is cosine function

$$\tilde{\alpha}_{r}(n) = \sum_{k=0}^{M-1} a(n-M+1+k) \exp[i2\pi(k-K)(r+d)/M] \; .$$

for the cosine apodizing function. The above expression defines the computation of fragment SDFT in the general case and is found through (6). $\{\alpha_{r}(n)\}$ may be restored through $\{\tilde{\alpha}_{r}(n)\}$ by using the well known properties of SDFT of complex-conjugate sequences ([2]).

The described above algorithms of local spectral analysis have been implemnted on the SM-4 mini-computer. The programs of recursive transform computation with respect to both coordinates are written in MACRO ASSEMBLER; computations were performed over floating-point numbers. The time of spectral analysis of 256x256 picture in the SDFT base is 0.15 h and 0.78 h for 8x8 and 16x16 fragments, respectively.

References

1. Yaroslawskij L.P. Adaptive Bildverarbeitungs-Techniken.- Bild und Ton, 1984, nol, s.25.
2. Yaroslavsky L.P. Digital Picture Processing. An Introduction.- Springer Verlag, Berlin-Heidelberg-New York-Tokyo, 1985.

FAST TRANSFORM METHODS OF IMAGE RECONSTRUCTION FROM PROJECTIONS AND THEIR PARALLEL IMPLEMENTATION ON SIMD-TYPE COMPUTER

Samuel MATEJ +

Abstract: A thorough analysis and simulation research of reconstruction errors and open problems in reconstruction from projections has been performed. The optimum reconstruction parameters has been obtained for Direct Fourier Inversion (DFI) method which proved to yield good image quality and was suitable for fast parallel implementation. Finally, parallel algorithm of DFI method for SIMD-architecture computer has been developed.

1. Introduction

The tomographic imaging methods represent one of the most topical and rapidly developing areas of image science. Despite of the simplicity and the mathematical exactness of the Direct Fourier Inversion method only the back projection (BP) methods are implemented in the contemporary CT-systems. The main reason for this was that of poor reconstruction quality having been attained by the DFI method while the BP method exhibited satisfactory results. The basic drawback of the BP method, namely a too high computational cost, was circumvented by implementation of a special hardware processor with the pipe-line architecture. In the recent years this method proved to be less convenient for new very fast tomographic methods, e.g. 3-D spatial methods, multiplanar methods, dynamical studies, even in case of a specialized hardware application. Moreover, the BP method does not suit also for nonexpensive tomographic applications which do not admit a special processor use. It seemed reasonable, therefore, to revalue the characteristics of the DFI projective algorithms which were more promising as to computational efficiency than the previous ones. The character of measured data structure in Magnetic Resonance (MR) tomography made such an analysis still more important.

We focused [4] our attention first of all on an investigation of discretization errors generation and suppression. Algorithms parameters have been analyzed taking the specific distinctions between MR and CT (X-ray) tomography into account. This made possible to carry out the thorough studies of the individual influences on image quality in the DFI method.

The optimum parameters have been determined in the sense of two criteria, i.e. the reconstructed image accuracy and computational speed. An important conclusion can be drawn from the computer simulation results: for comparable reconstructed image accuracy in case of both projective methods a dramatic computational advantage has been reached (especially for large matrix size) for the DFI method.

+ Institute of Measurement and Measuring Technique of Electro-Physical Research Centre Slovak Academy of Sciences, Dúbravská cesta 9, 842 19 Bratislava, Czechoslovakia

It can be assumed that the new generation of tomographic systems will comprise a more sofisticated image processing system based on parallel architecture principle. This computational capability will provide at the same time an opportunity to use for very fast tomographic imaging the DFI method. A SIMD-like parallel architecture has been found [4] to be the most convenient for DFI algorithms implementation. Among three basic steps of DFI method (filtration, interpolation, FFT) the parallel interpolation of a polar raster in the Fourier spectral domain is the most complicated and less investigated one. Therefore a special attention was focused on a parallel algorithm design of this part of the DFI method.

The similar philosophy and the parallel algorithm structure seem to be applicable in the very fast MR-imaging methods based on other nonlinear spatial raster scanning.

2. The basis of the Fourier Inversion method

Generally the task of tomographic reconstruction is to find the unknown function f(x,y) (representing some quantitative characteristics-physical features of examined objects) on the basis of measured indirect data. The measured data in tomographic methods are represented by projections $p(s;\theta)$ - the integral values of the function f(x,y) along the lines perpendicular to the projection lines - with angle θ and distance s from the origin - (see fig.1), or by their Fourier transformation $P(S;\theta) = \mathcal{F}[p(s;\theta)]$ (in MR-tomography). The projections can be described [1] by Radon

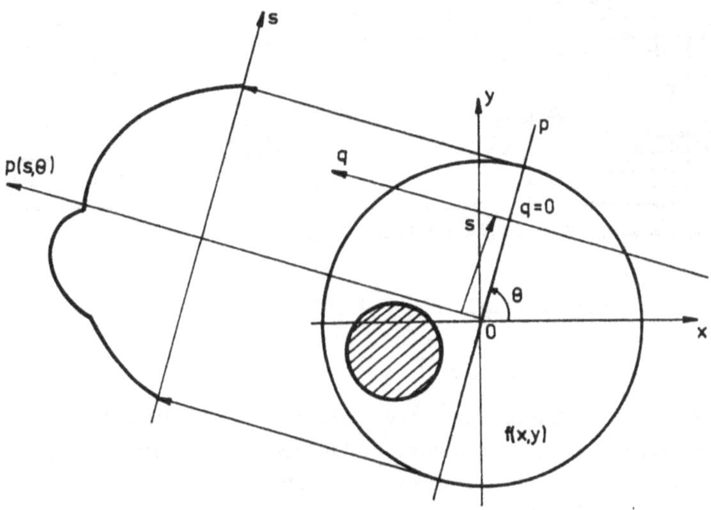

Fig. 1 The Radon transformation geometry

transformation:

$$p(s;\theta) = \mathcal{R}[f(x,y)] = \int_{-\infty}^{\infty} f(s.\cos\theta - q.\sin\theta,\ s.\sin\theta + q.\cos\theta)\,dq\ . \qquad /1/$$

The objective of image reconstruction from projections is then to find the inverse Radon transformation. In the continuous case there are several equivalent formulas of it. The formula of examined DFI method is based on the projection theorem:

$$[\,\mathcal{F}_y \mathcal{R}f\,](S;\theta) = [\,\mathcal{F}_2 f\,](S.\cos\theta,\ S.\sin\theta)\ , \qquad /2/$$

where $\mathcal{F}_y$ is the 1-D Fourier transform (FT) operator of projections and $\mathcal{F}_2$ is the two dimensional Fourier transform (2-D FT) operator. Further, from /2/ we can derive the DFI method formula (for CT and MR-tomography):

$$f(x,y) = [\mathcal{F}_2^{-1}\,\mathcal{F}_y\,\mathcal{R}]\ f(x,y) = [\,\mathcal{F}_2^{-1}\ \mathcal{F}_y\,]\,p(s;\theta) = \mathcal{F}_2^{-1}\ P(S;\theta)\ . \qquad /3/$$

In actual discrete reconstruction problems the transform methods are not more equivalent because of the discretization errors and not exact approximations of transform formulas. The main goal of research of image reconstruction methods is to find method of reconstruction from the measured discrete projection data yielding the best estimation $\hat{f}(x,y)$ of the $f(x,y)$. It was found out that in the DFI method the error of nonexact interpolation from the polar raster of Fourier transformed projections to the rectangular raster of Fourier spectral data has the greatest influence on the reconstruction quality [5, 7]. The interpolation procedure is evoked by the fact that in practical cases only the finite number of polar raster points is available while the 2D FT algorithms require the data given on the rectangular raster. The most succesfull way to reach a good reconstructed image quality in the DFI method is to solve the problem of sufficiently precise interpolation. The basic steps of practical DFI method algorithm are illustrated in fig. 2.

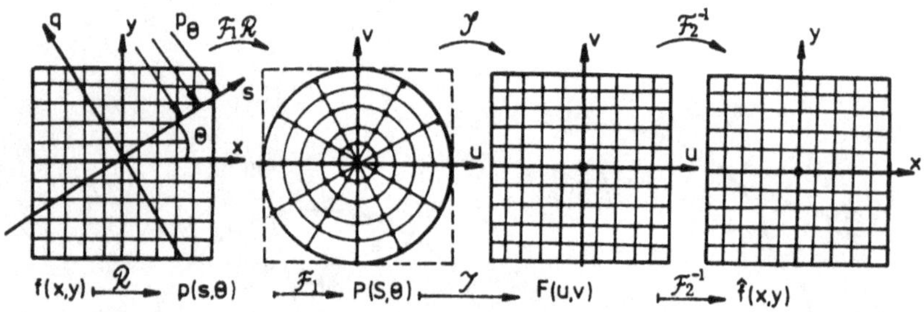

Fig. 2 Basic steps of DFI method algorithm

3. Research of DFI method characteristics

There are two main sources of reconstruction errors in discrete mathematical model: not exact discrete approximation of continuous reconstruction formulas and the inevitabely finite number of discrete samples. Although, there exist some ideas concerning the possibilities of suppresion reconstruction errors of DFI method (e.g [2, 5, 6, 7]), to perform a complex analysis of mutual relations of these errors, still remains of great interest. A MODUlar programming System for SImulation of Projective Reconstruction methods - MODUSIRP has been developed. This system makes possible to carry out computer simulation of various approaches to error suppresion in distinct reconstructive methods. The system enables also simulation research of various discretization schemes, interpolation and filtering techniques. It comprises the image projection generators from mathematical phantoms and real Byte images, the simulated noise generators and the means of reconstruction quality evaluation (more about

MODUSIRP see [3,4]). The results of theoretical analysis and computer experiments [4] can be summarized as follows.

a) Errors caused by the spatial domain discretization (this is not the case in MR-tomography)

The bounded step of projections discretization leads to the periodic repeating of their Fourier spectrum which causes in turn an aliasing error - superposition of the border lobes from neighbouring periods. This error can not be totally eliminated because as is known the Fourier spectrum support of the reconstructed function with finite support is infinite (see information theory). The error can be suppressed to the neglected level by a sufficiently small choice of the discretization step of projections (Fourier spectrum period is rising) and by an appropriate projection data filtration in the spatial domain before Fourier transformation (in CT-tomography it is done by the finite dimensions of detectors).

b) Errors caused by the Fourier spectrum domain discretization

These errors are present in methods using Fourier spectrum domain. For application of convolution BP method in MR-tomography it is necessary to consider a new type of error therefore, inspite of its use in CT-tomography.
Incorrect choice of the discretization step can bring an aliasing error in the space domain. However, the correct choice of the step is not a problem due to the finite support of real functions to be reconstructed. The more important fact consists in the finite number of Fourier spectrum samples (finite extension). The effect of it is the same as the Fourier spectrum truncation or filtration with a rectangular window [5]. This is the source of oscillating artifacts (especially near the objects with a large contrast,e.g. near the scull) which appear in the reconstructed image. If the discretization step of reconstructed image is fixed, the only possible way to suppress these oscillations is to use a spectral filter with smooth tapers at the ends at the expense of a slightly smoothing the edges of image. Computer experiments showed the filter of cos type to be the most convenient. There is also another kind of artifacts caused by the Fourier domain discretization. It has been found out that they can be corrected by enlargement of the 2-D FT dimension (this can be done by enlargement of the spatial domain by zero values via a reduction of the discretization step in Fourier spectral domain).
The experiments showed approximately 25% enlargement of the 2-D FT dimension is sufficient.

Computer simulation implies that reconstruction error suppression described till now is necessary in those cases the objects to be recognized in a reconstructed image exhibited the local contrast not greater than 1-10 % of their sorrounding object values (e.g. in the cases of brain examination). However, when this is not the case, i.e. the objects exhibited the contrast greater than 10 %, the above mentioned correction is not any longer necessary (the artifact patterns are masked by the contrast in image).

Finally, we can point out here that the Fourier domain discretization has a great influence on the interpolation accuracy reviewed in the next paragraph.

c) Errors caused by interpolation of cartesian raster (from polar raster)

The quality of reconstructed image in DFI method depends mostly on the accuracy of interpolation in the Fourier spectrum domain [5, 6, 7] and experiments confirmed the great dependence of reconstructed image quality on radial projections enlargement (reduction of the radial discretization step in Fourier domain). It was also found, that nonexact interpolation can enlarge some influences described earlier, or can together with them produce some new artifacts. Computer simulation showed that a very good image quality may be obtained by bilinear interpolation, which can be very effectively algorithmized. This requires the necessary enlargement of projection vector to be 4 - 8 .

d) Errors caused by angular discretization

Theoretically sufficient number of projections (views) for a good reconstruction equals $\pi N/2$ (N denotes the dimension of rectangular raster) [7] . The experiments showed that the practically sufficient number is dependent on a reconstructed image character and may be chosen smaller in many cases.

e) Measurement noise

We considered the gaussian white noise which is a good approximation of actual conditions. The computer experiments showed that the acceptable values of noise variance in measured data are:

- comparable with their objects local contrast for the objects of dimension of several pixels,
- 2-3 times larger than their contrast for larger objects (their dimensions are minimally ten times larger than the pixels distance).

The optimum parameters for reconstruction do not change in the case of acceptable values of projections noise. If the larger values of noise are present, it is necessary to suppress them by accumalation measurement process or computational process e.g. by increasing the number of Fourier spectrum samples for the 2-D FT. The experiments confirmed that the noise suppression in this case is proportional to the Fourier Transform dimension enlargement. The influence of the other computational steps of image reconstruction methods on the noise suppression has been studied in more details.

The optimum parameters of image reconstruction by the DFI method derived for bilinear interpolation are as follows:

- the cos type of filtration, the 25 % enlargement of the dimension of the 2-D FT, the enlargement factor of projections equal to 4-8 for objects with the small contrast (several percents),
- no filtration, no enlargement of the 2-D FT dimension, the enlargement factor of projections equal to 4 for objects with large contrast (exceeding 10 %).

In comparison with the BP convolution method the quality of images reconstructed by the DFI method with these parameters was very good. The reconstruction time saving, however, proved to be considerable. In the case of practical implementation of reconstruction methods on minicomputer SM 52-11 (PDP-family) for 102x102 pixels images the factor of computational time reduction was 17 in the case of CT-tomography and 43 in the case of MR-tomography.

4. A parallel algorithm design for the DFI method

The theoretical and simulation research [4] of the DFI method that was carried out indicates that this method is a promising candidate for ultra fast tomographic imaging implementation. The detailed analysis of its crucial points leds to the decision to favour the SIMD-type architecture.

4.1 The parallelization of the basic steps of DFI method

The Fourier transformation (FT) of projections represents the first step of the DFI method in CT-tomography. The parallel SIMD architecture fits for the FT computation very well. The projections enter succesively the parallel processor, where the FT is being performed. As the parallel structure of the internal data is not suitable for the interpolation algorithm, the data have to be sorted outside the parallel processor. It is useful to include the filtration step too. In MR-tomography the FT of projections is measured directly and therefore it is more effective to perform the filtration together with the interpolation computations.

A parallelization of algorithms for the rectangular spectral raster interpolation is not a trivial task. Since its solution has not been encountered yet in the literature we shall describe the basic features of a parallel bilinear interpolation intended for the implemantation in a SIMD-type computer (for practical algorithm design see [4]). The similar philosophy can be used also for other interpolation types.

The inverse 2-D Fourier transform (2-D IFT) is the last step of the DFI algorithm. As the parallel implementation of the FFT is well known the latter task (2-D IFT) can be accomplished in a SIMD-type computer without any problem.

4.2 The bilinear parallel algorithm design

The angular sectors of a polar raster in the spectral domain are being processed sequentially during the interpolation procedure. The parallel approach is applied to computations within the single sectors only.

Let us consider the bilinear interpolation with the geometry depicted in fig. 3. In this case the following formulas hold:

$$F_A = D_d F_2 + D_h F_1 , \quad F_B = D_d F_3 + D_h F_4 , \quad F = U_d F_B + U_h F_A , \qquad /4/$$

where F_i , $i = 1, 2, 3, 4$ denote the function values in the points of polar (P) raster; F_A , F_B denote the linearly interpolated values in the points A, B; F represents the interpolated function value in the rectangular (R) raster point; and D_d, D_h, U_d, U_h are relative distance measures.

It is useful to point out that the polar raster border circle is situated inside the rectangular raster border square (see fig. 2). We set the R raster points outside the circle to be zero. The parallel algorithm is devided into these three stages.

In the I. stage. the points of R raster involved in the given angular sector are searched. The searching is performed parallelly along the R raster lines, which are perpendicular to the X-axis. Having finished this stage, for every line the number of its points situated in the given sector and the smallest Y coordinate are available. These data serve as the basis for points coordinates determination which are necessary in the next stage.

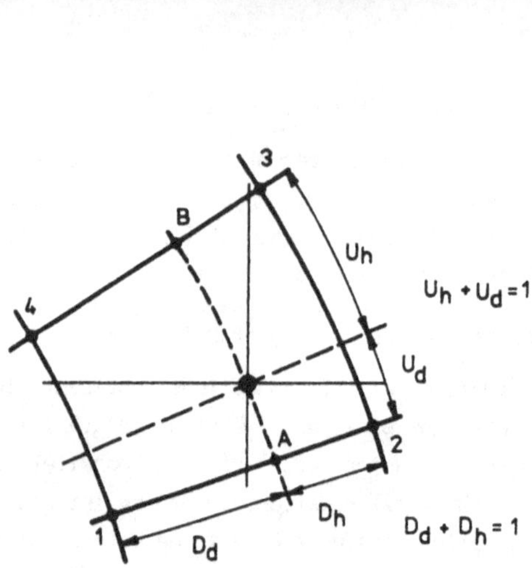

Fig. 3 Bilinear interpolation geometry

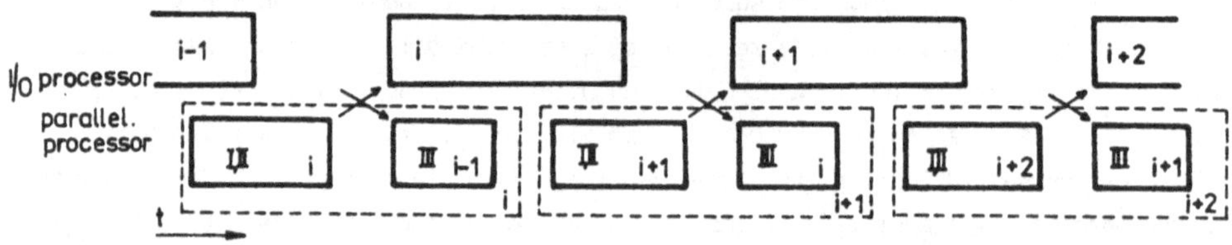

Fig. 4 Pipe-lining the work of I/O processor and parallel processor

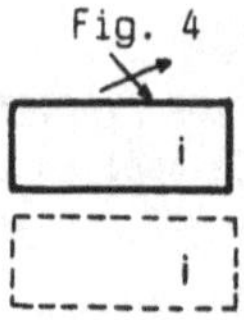

- indicates the data exchange
- indicates the stage of computations for the ith sector
- indicates the ith cycle of global interpolation algorithm for all sectors

In the II. stage, all coefficients and data necessary for interpolation are computed parallelly for all R raster points from the given sector. The calculations start with the polar coordinates of these points, further the coefficients D_d, D_h, U_d, U_h and the corresponding indexes of FT projection vectors (which enclose the given angular sector) are calculated.

The obtained data are moved to I/O processor. The I/O processor is preparing the values of the nearest P raster points for each R raster point from a given sector. All necessary data (P raster values and coefficients) are arranged to have a convenient structure for final interpolation calculation. After arranging they are moved to the parallel processor back again. To make the whole procedure more effective it is useful to pipe-line the work of parallel processor and I/O processor. This can be accomplished by computing the data of the I. and II. stage for the $i+1^{th}$ angular sector while the III. stage data are computed for the ith sector (see fig. 4). Therefore, the I/O processor can prepare the data for the III. stage in advance parallelly to the data processing in parallel processor. This speeds up the computations because the parallel processor need not wait while the I/O processor is busy in preparing the data.

Finally, in the III. stage, the main interpolation computations (formulas /4/) are performed parallelly for points of the given sector. In the case of MR-tomography the filtration is also performed, whereby the filter values corresponding to the rectangular raster points are calculated on the basis of their R raster coordinates.

5. Conclusion

The Direct Fourier Inversion method of reconstruction from projections has been analyzed. There has been found the ways of reconstruction errors suppression in various reconstruction situations. The reconstructed image quality was comparable with the quality of convolution back projection method, while the reconstruction times were drastically decreased. There have been designed the parallel algorithm for SIMD-type computer. On the basis of our research we recommend the DFI method for use in the fast spatial and dynamic tomographic applications. We recommend it also in the nonexpensive applications where the specialized computer can not be used.

References

[1] Herman, G.T.: Image reconstruction from projections - Implementation and applications. Springer - Verlag, New York 1979.

[2] Herman, G.T.: Image reconstruction from projections - The fundamentals of computerized tomography. Academic press, New York 1980.

[3] Matej, S., Bajla, I.: A software system for simulation of image reconstruction from projections in MR-tomography. To be published in proc. of Conference on Computer Analysis of Images and Patterns CAIP'87, Berlin, 2.-4. Sept. 1987.

[4] Matej, S.: A contribution to projection reconstruction methods in tomographic systems.Slovak Academy of Sciences, Bratislava 1987, Ph. D. Thesis.

[5] Mersereau, R.M.: Direct Fourier transform techniques in 3-D image reconstruction. Comput. Biol. Med. 6 (1976), pp. 247-258.

[6] Niki, N., R.T. Mizutani, Y. Takahashi, T. Inouye: A high-speed computerized tomography image reconstruction using direct two-dimensional Fourier transform method. Scripta Electronica Japonica III 14 (1983) 3, pp. 56-65.

[7] Stark, H., J.W. Woods, I. Paul, R. Hingorani: An investigation of computerized tomography by direct Fourier inversion and optimum interpolation. IEEE Trans. Biomed. Eng. BME-28 (1981) 7, pp. 496-505.

ERROR - FREE DECONVOLUTION AND ITS APPLICATION IN SIGNAL PROCESSING

Miroslav Morháč [1)]

1. Introduction

In the past years increased attention has been paid to the problems of deconvolution [2-5]. Very frequently the advantageous property of factorization of two signals convolution to the product of their Fourier coefficients is utilized [1,7,8]. Since the Fourier transform cannot be computed with absolute precision, the number theoretical transforms were defined with the aim to perform the fast error-free convolution of signals. The problem of precision is much more critical mainly in the inverse operation, i.e. in deconvolution.

The error-free algorithm removes rounding-off errors which are unavoidable when using Fourier transform. It can be used to enhance the resolution in one and two parameter signals.

2. General considerations

The dynamic behaviour of a linear system can be described by means of impulse-response function. Let us denote the impulse response of such a discrete system by h(n) and the input by x(n). Then the periodical convolution of finite sets x(n) and h(n) is defined by

$$y(n) = \sum_{m=0}^{N-1} x(m)h(n-m) \quad , \; n = 0,1,\ldots,N-1 \tag{1}$$

or for two dimensional system by

$$y(n_1,n_2) = \sum_{m_1=0}^{N-1} \sum_{m_2=0}^{N-1} x(m_1,m_2)\, h(n_1-m_1,n_2-m_2), \; n_1,n_2 = 0,1,\ldots,N-1 \tag{2}$$

where the sets x(n), h(n) and $x(n_1,n_2)$, $h(n_1,n_2)$ are periodically extended with the period N, i.e. their indices are calculated modulo N. Then (1) can be written as

$$\begin{bmatrix} y(0) \\ y(1) \\ \vdots \\ y(N-1) \end{bmatrix} = \begin{bmatrix} h(0) , & h(N-1) & \ldots & h(1) \\ h(1) , & h(0) & \ldots & h(2) \\ \vdots & \vdots & & \vdots \\ h(N-1) , & h(N-2) , & \ldots & h(0) \end{bmatrix} \cdot \begin{bmatrix} x(0) \\ x(1) \\ \vdots \\ x(N-1) \end{bmatrix} \tag{3}$$

1) Institute of Physics, Electro-Physical Research Centre, Slovak Academy of Sciences, 842 28 Bratislava, Dúbravská 9, Czechoslovakia

or

$$\bar{y} = \bar{\bar{h}}_1 \cdot \bar{x} \tag{4}$$

Let us create the submatrix H_i by cyclic shifting of i-th row of response matrix $h(n_1,n_2)$ as

$$H_i = \begin{bmatrix} h(i,0) & , h(i,N-1) , & \cdots & h(i,1) \\ h(i,1) & , h(i,0) \quad , & \cdots & h(i,2) \\ \vdots & \vdots & & \vdots \\ h(i,N-1) , & h(i,N-2) , & \cdots & h(i,0) \end{bmatrix} \tag{5}$$

Then it holds

$$\begin{bmatrix} y(0,0) \\ \vdots \\ y(0,N-1) \\ y(1,0) \\ \vdots \\ y(N-1,N-1) \end{bmatrix} = \begin{bmatrix} H_0 & , H_{N-1} & , \dots , & H_1 \\ H_1 & , H_0 & , \dots , & H_2 \\ \vdots & \vdots & & \vdots \\ H_{N-1} & , H_{N-2} & , \dots , & H_0 \end{bmatrix} \cdot \begin{bmatrix} x(0,0) \\ \vdots \\ x(0,N-1) \\ x(1,0) \\ \vdots \\ x(N-1,N-1) \end{bmatrix} \tag{6}$$

or

$$\bar{y} = \bar{\bar{h}}_2 \cdot \bar{x} \tag{7}$$

The submatrices H_0 to H_{N-1} in (6) are mutually periodically shifted.

Both one and two-dimensional systems can be solved by several methods :

a. Direct solution of matrix equation (4) or (7) .

This process has the following properties :

- it requires the caculation of the inverse matrix. Of course it is a problem mainly for large systems
- there inevitable arise rounding-off errors due to the finite length of the computer word

b. Solution using Fourier transform. From equation (1) or (2) we obtain

$$F_i[y(n)] = F_i[h(n)] \cdot F_i[x(n)] \quad , \quad i = 0,1, \dots , N-1 \tag{8}$$

or

$$F_{i_1,i_2}[y(n_1,n_2)] = F_{i_1,i_2}[h(n_1,n_2)] \cdot F_{i_1,i_2}[x(n_1,n_2)] \tag{9}$$

$$i_1,i_2 = 0,1, \dots , N-1$$

Then it holds that

$$F_i[x(n)] = F_i[y(n)] \, F_i[h(n)]^{-1} \tag{10}$$

or

$$F_{i_1,i_2}[x(n_1,n_2)] = F_{i_1,i_2}[y(n_1,n_2)] F_{i_1,i_2}[h(n_1,n_2)]^{-1} \tag{11}$$

By inverse transformations (one or two dimensional) of equations (10),

(11) we get the solution with the following properties

- it does not require the calculation of the inverse matrix
- there arise rounding-off errors since the transformed values are real numbers. The mutual division of these rounded-off numbers in (10), (11) may critically affect the result.

c. Solution using Fermat transform

The Fermat transform also factorizes the periodical convolution of signals to the modulo product of their coefficients. Analogically to (8), (9) it can be written that

$$R_i\left[y(m)\right] = R_i\left[h(n)\right] R_i\left[x(n)\right] (\mathrm{mod}\ M) \quad , i = 0,1,\ldots,N-1 \tag{12}$$

or

$$R_{i_1,i_2}\left[y(n_1,n_2)\right] = R_{i_1,i_2}\left[h(n_1,n_2)\right] R_{i_1,i_2}\left[x(n_1,n_2)\right] (\mathrm{mod}\ M) \tag{13}$$

$$i_1, i_2 = 0,1,\ldots,N-1$$

and M is the Fermat number defined as

$$M = F_t = 2^{2^t} + 1 \quad t = 0,1,2,\ldots \tag{14}$$

The properties of the solution are :

- it does not require the calculation of inverse matrix
- there do not arise rounding-off errors due to the integer operations executed in modular arithmetic [9].

3. Algorithm of two dimensional deconvolution using Fermat transform

Two dimensional Fermat transform pair can be expressed as

$$X(a_1,a_2) = \left[\sum_{n_1=0}^{N-1} \sum_{n_2=0}^{N-1} \alpha^{a_1 n_1} \cdot \alpha^{a_2 n_2} \cdot x(n_1,n_2)\right] (\mathrm{mod}\ M) \tag{15}$$

respectively

$$x(n_1,n_2) = \left[N^{-2} \sum_{a_1=0}^{N-1} \sum_{a_2=0}^{N-1} \alpha^{-a_1 n_1}\ \alpha^{-a_2 n_2}\ X(a_1,a_2)\right] (\mathrm{mod}\ M) \tag{16}$$

where n_1, n_2, a_1, a_2 = 0,1,...,N-1; N is a power of 2 and there holds

$$\alpha^N (\mathrm{mod}\ M) = 1 \tag{17}$$

Similarly the same transform pair can be written in matrix notation as

$$X = \left[T \cdot x \cdot T\right] (\mathrm{mod}\ M) \tag{18}$$

resp.

$$x = \left[T^{-1} \cdot X \cdot T^{-1}\right] \pmod M \tag{19}$$

where T, T^{-1} are Fermat resp. inverse Fermat transform matrices. The error-free one dimensional algorithm was derived and published in [6]. According to this derivation the two-dimensional error-free deconvolution algorithm can be described as follows

1. Calculate the matrix of the transformed values of the two-dimensional response function h

$$H = T \cdot h \cdot T \pmod M \tag{20}$$

2. Calculate the matrix H^{-1} such that for all their elements $H^{-1}(i_1,i_2)$ there holds

$$H(i_1,i_2) \cdot H^{-1}(i_1,i_2) \pmod M = 1 \quad i_1,i_2 = 0,1,\dots, N-1 \tag{21}$$

Using the matrix H from eq.(20) we can write for the determinant of system (7)

$$D \pmod M = D_M = \left[\prod_{i_1=0}^{N-1} \prod_{i_2=0}^{N-1} H(i_1,i_2)\right] \pmod M \tag{22}$$

3. Let $ym_0(n_1,n_2) = y(n_1,n_2) \quad n_1,n_2 = 0,1,\dots,N-1$
4. Calculate the transform of ym_0

$$YM_0 = \left[T \cdot ym_0 \cdot T\right] \pmod M \tag{23}$$

5. For $i_1,i_2 = 0,1,\dots,N-1$ calculate

$$X_0(i_1,i_2) = \left[H^{-1}(i_1,i_2) \cdot YM_0(i_1,i_2)\right] \pmod M \tag{24}$$

6. The inverse transform of X_0 is

$$x_M = \left[T^{-1} \cdot X_0 \cdot T^{-1}\right] \pmod M = \frac{D}{D} x_M \pmod M = \frac{D_M}{D} x_M \pmod M = \frac{x_0}{D} \tag{25}$$

7. Let $x(i_1,i_2) = x_0(i_1,i_2) = D_M X_M(i_1,i_2) \pmod M$ (26)

$$i_1,i_2 = 0,1,\dots, N-1$$

8. According to (6), (7) calculate the vector

$$\overline{y_0'} = \overline{\overline{h_2}}\ \overline{x_0} \tag{27}$$

form the matrix $y_0^{\bullet}$ and then calculate matrices

$$y_1 = \frac{1}{M}\left(ym_0 \cdot D - y_0^{\bullet}\right) \tag{28}$$

$$ym_1 = y_1 \pmod M \tag{29}$$

9. Let j = 1
10. The transform of ym_j is

$$YM_j = \left[T \cdot ym_j \cdot T\right] \pmod M \tag{30}$$

11. For i_1, i_2 = 0,1,...,N-1 calculate

$$X_j(i_1,i_2) = \left[H^{-1}(i_1,i_2) \cdot YM_j(i_1,i_2)\right] (\mathrm{mod}\ M) \tag{31}$$

12. The inverse transform of X_j

$$x_j = \left[T^{-1} \cdot X_j \cdot T^{-1}\right] (\mathrm{mod}\ M) \tag{32}$$

13. For i_1, i_2 = 0,1,...,N-1 calculate

$$x(i_1,i_2) = x_j(i_1,i_2) + x_j(i_1,i_2) \cdot M^j \tag{33}$$

14. According to (6), (7) calculate the vector

$$\bar{y}_j^{\bullet} = \bar{\bar{h}}_2 \cdot \bar{x}_j \tag{34}$$

form the matrix $y_j^{\bullet}$ and then calculate the matrices

$$y_{j+1} = \frac{1}{M}(y_j - y_j^{\bullet}) \tag{35}$$

$$ym_{j+1} = y_{j+1} (\mathrm{mod}\ M) \tag{36}$$

15. If for all i_1, i_2= 0,1,...,N-1 $y_{j+1}(i_1,i_2) = 0$, then finish the calculation. If not, increment j and repeat from point 10 on.
16. $x(i_1,i_2) = x(i_1,i_2) / D \quad i_1,i_2 = 0,1,\ldots,N-1$ (37)

Well known Euclidian algorithm can be used for calculation of needed inverse number (see eq. 21) in the framework of modular arithmetic. One version of such algorithm is described in [7].

The error-free algorithm also requires the value of determinant in (28), (37). In [6] the algorithm for precise calculation of the determinant of one-dimensional convolution system was introduced. The algorithm to calculate the determinant of matrix for two-dimensional case (see eq. (6) and (7)) was found and described in [7].

4. Applications

The described method of deconvolution was used to the resolution enhancement in processing of spectroscopic data. The algorithms were applied to data given in fig.1 and 3.

Fig. 1 shows spectrum consisting of 4 peaks located very closely to each other (positions 86, 89, 93, 97 ; heights 150, 100, 50, 200). Looking at this figure we cannot state the existence of second and third peak in the middle of the spectrum. Fig. 2 shows the spectrum after deconvolution.

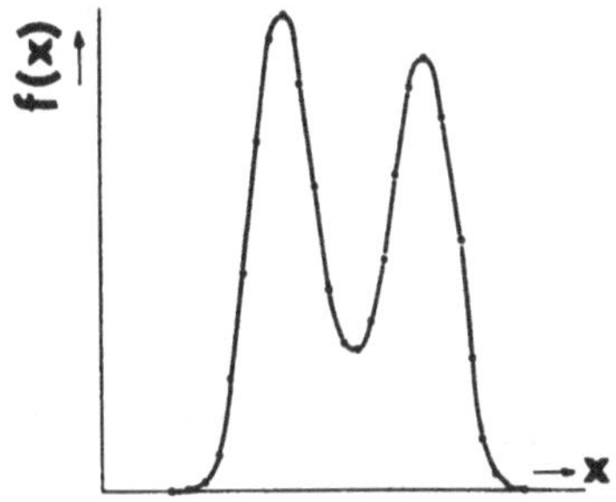

Fig. 1

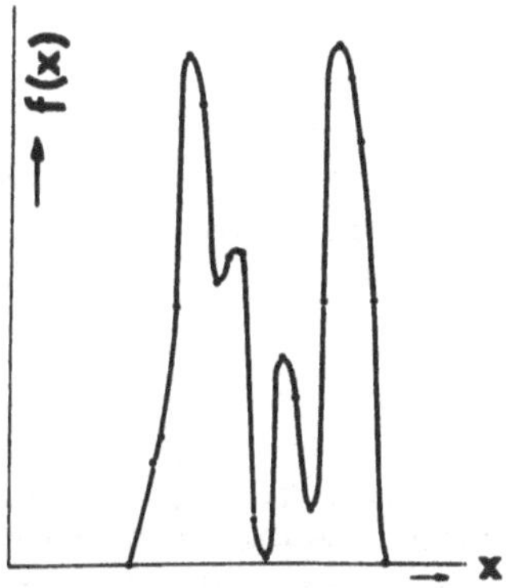

Fig. 2

Similarly the two-dimensional deconvolution algorithm described in part 3 of this paper was verified. Fig. 3 shows the spectrum before deconvolution (positions $x_1 = 7, y_1 = 9$, $x_2 = 10$, $y_2 = 8$, $x_3 = 8$, $y_3 = 5$; heights 200, 150, 50) and fig. 4 shows deconvoluted spectrum with evidently better resolution.

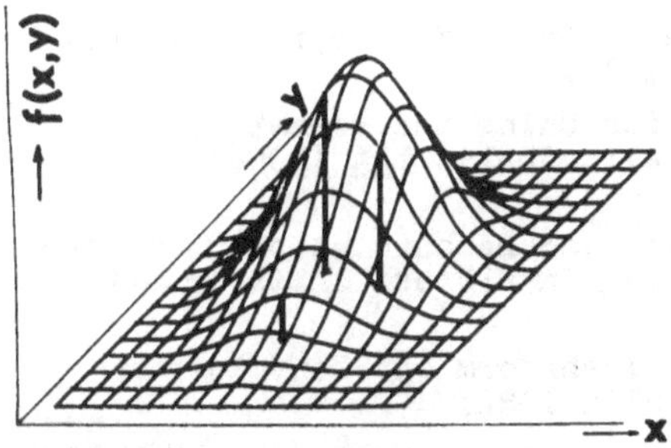

Fig. 3

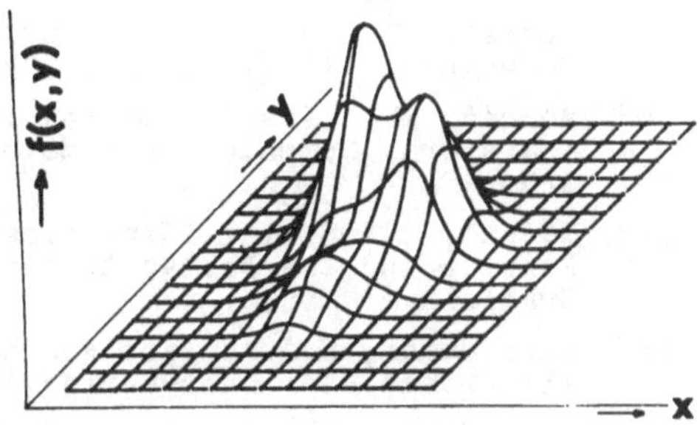

Fig. 4

The algorithms were implemented for a TPA-70 minicomputer (Hungarian production). The modulus $M = F_4 = 65537$ and coefficient $\alpha = 3$ were chosen in order to permit the length of the transformed vector or matrix up to 65536.

5. Conclusion

The method described in the paper shows that Fermat transform can be successfully used also for the two-dimensional deconvolution. The main point of the usage of modular arithmetic is an elimination of the rounding-off errors. On the other hand commonly used Fourier transform gives a solution with oscillations.

Unfortunately, the algorithm is rather time-consuming since the calculation process must be repeated several times. But there is a possibility to eliminate this disadvantage using specialized processor of the fast Fermat transform.

6. Literature

[1] Ahmed,N., Rao, K.R. : Orthogonal Transforms for Digital Signal Processing. Springer-Verlag, Berlin 1975.

[2] Goutte, R., Prost, R., Georges, A. : Déconvolution numérique avec prolongement spectral. Applications aux signaux et aux images. Analusis 8 (1980) 1, 6-15.

[3] Grabaric, B.S., O'Halloran, R.J., Smith, D.E. : Resolution Enhancement of A.C. Polarographics Peaks by Deconvolution Using Fast Fourier Transform. Analytica Chimica Acta 133 (1981), 349- 358.

[4] Kennet, T.J., Prestwich, W.V.,: Incremental Deconvolution I. Algorithm Development and Assessement. Nuclear Instruments and Methods 203 (1982), 317-327.

[5] Lacoste, L.J.B. : Deconvolution by Succesive Approximations. Geophysics 47 (1982) 12, 1724-1730.

[6] Morháč, M. : Precise Deconvolution Using the Fermat Number Transform. Computers and Math. with Appls. 12A (1986) 3, 319-329.

[7] Morháč, M.: System Identification and Deconvolution Using the Fourier and the Fermat transforms. In Slovak . Dissertation, Bratislava 1983.

[8] Nussbaumer, H.J. : Fast Fourier Transform and Convolution Algorithms. Springer-Verlag, Berlin 1981.

[9] Ore, O. : Number Theory and Its History. McGraw-Hill, New York 1948.

Integral Invariant Description of Grey-Scale Images

A.E.Krasnov, A.K.Glotov, V.F.Starodubtsev [1)]

Introduction

In certain applikations of the automatic monitoring one has to generate image description that are independent of unknown initial conditions of the object in question. A most complicated and most common is the case where all the information on object state is contained in the degrees of grey of its image. In this connection, the problem of constructing an integral invariant image description cannot be regarded as the new. When one comes across an individual object or a group of objects making up a complex scene, the problem may by of interest by itself or be the basic element of a more general, e.g. structural, description. Mathematically, construction of the integral description involves calculation of various functionals of two-dimensional functions which may be exemplified by the well known methods of moment, spectral or statistical invariants [1,2,3]. The ever increasing amount of methods for separation of image features attests the importance of the problem and, at the same time, points to the lack of a unique approach to it [4]. The lack of generality is the common disadvantage of these methods because they depend on the nature of image and disregard the physical properties of the image as the subject of inquire. As the result, degenerate descriptions are obtained where similar descriptions correspond to structurally different images, and the causes of degeneracy and the ways to its elimination are not clear.

This paper aims at an integral invariant method of image description that is free of the above demerits and emphasizes a description invariant to translation-and-scale image transformations. Image is primarily regarded here not as a function but as a physical object or the result of interaction of certain physical processes such as spatial propagation of electromagnetic oscillations or waves and detection of physical values related to them. Owing to the common nature of the dynamic processes of electromagnetic wave propagation, the method suggested here may be used for image description in the optical, radio-frequency or X-ray bands. In each of the bands there are, however, specific features of wave detection. In this connection, the paper considers the promising digital, analogue and optical hardware for implementation of the proposed method in different applications.

[1)] The authors are with the Institute of Control Sciences, 65, Profsoyuznaya st., Moscow, 117342, USSR.

Qualitative description of image. Phase portrait of the image spatial structure.

Let us consider a scalar electromagnetic signal corresponding to the intensity of electromagnetic field propagating from the object under consideration to the detector. The signal is described in mathematical terms as a real function $E(x,y,z,t)$ of space (x,y,z) and time (t) coordinates. The interaction of the field and detector including operations of measurement and computation results in the image $I(x,y,t)$, a positive definite function giving mathematically adequate spatial structure of the electromagnetic signal observed at the aperture of the detector. It goes without saying that in the process of image generation the physical properties of both signal and detection are taken into account in order to render as much information contained in the signal as possible. In the final image, however, the usefui qualitative properties of the signal, interrelations between its individual components are reduced. Description of their relations is the integral image description. This interrelation may be established by taking into account the space-time relations between the components of electromagnetic signal E or the physical process corresponding this signal. This space-time relation is known to be expressible according to the laws of the classical electrodynamics as the wave equation $c^2 \nabla^2 E = \partial^2 E/\partial t^2$ that is satisfied by field E in the free space [5]. In practice, at image generation one observes quasi-stationary electromagnetic fields $E(x,y,z,t) = f_t(x,y,z) \cos(\omega t)$ having the variation speed of envelope f much smaller then the mean cyclic frequency ω of the electromagnetic signal carrier. The envelope or the "video" signal f_t satisfies the wave equation $\nabla^2 f_t + k^2 f_t = 0$ where k is the wave number corresponding to ω and light speed c in vacuum ($k = \omega c$). Superposition of flat waves propagating from the object to detector is the real part of the equation solution having a profound physical sense [6] :

$$f_t(x,y,z) = \sum_{kx} \sum_{ky} \sum_{kz} R_{kx,ky,kz}(t) \exp(-i(k_x x+k_y y+k_z z)),$$

where $k_x^2+k_y^2+k_z^2 = k^2$ and the amplitudes $\{R\}$ of flat waves depend on the boundary conditions as defined by the object and the distance to the detector aperture. Let us group the terms of the solution so as to explicitly separate the harmonic dependence of lines and omit the coordinate $z=Z$ of the flat detector aperture:

$$f_t(x,y) = f_t(x,y,Z) = \sum_{kx} R_{kx}(y,t) \exp(-ik_x x)$$

Thus, the video signal of any line y in the detector aperture plane is the superposition of harmonic processes $q_{kx}(x)$ with different spectral weights R_{kx} satisfying each the second-order equation $d^2 q_{kx}/dx^2+k_x^2 q_{kx}=0$.

On the whole, the harmonic process may be described by introducing the phase plane $(q_{kx}, p_{kx} = dq_{kx}/dx)$ [7]. The affix trajectory over the phase plane will be the process phase portrait which is ellipse defined by a Hamiltonian of the process $H_{kx} = k_x^2 R_{kx}^* R_{kx} = k_x^2(q_{kx}^2 + p_{kx}^2/k_x^2)$ where R^*R is constant value for each spectral component of signal. The video signal $f_t(x,y)$ may be qualitatively characterized by the totality of these phase portraits for various spectral components k_x of various lines y. However, there are two reasons why the totality is inconvenient for video signal description. First, the dimensionality of the description is great, and, second, the description is not invariant to scale transformations of the signal which is due to the fact that the affix trajectory of each harmonic signal is explicitly dependent on the wave number k_x. Transform the coordinates q_{kx} and p_{kx} of the multi-dimensional phase space describing the totality of harmonic processes forming the video signal: $\tilde{q}_{kx} = q_{kx}$, $\tilde{p}_{kx} = p_{kx}/k_x$. The phase portraits $R_{kx}^* R_{kx} = \tilde{q}_{kx}^2 + \tilde{p}_{kx}^2$ of harmonic processes are independent of their frequencies and, thus, are invariant to variations of signal scale.

In order to reduce dimensionality of the description project all the "current" coordinates of phase trajectories on the corresponding axes $\tilde{q}$ and $\tilde{p}$ in the same phase plane and sum them up with the weights $R_{kx}(y,t)$, thus forming the phase space with generalized coordinates Q and P:

$$Q_t(x,y) = \sum_{kx} R_{kx}(y,t)\, \tilde{q}_{kx}(x) = \sum_{kx} R_{kx} \exp(-ik_x x) = f_t(x,y),$$

$$P_t(x,y) = \sum_{kx} R_{kx}(y,t)\, \tilde{p}_{kx}(x) = \sum_{kx} R_{kx}\, (dq_{kx}/dx)/k_x .$$

The latter transformation is exactly the Hilbert transformation of the signal f_t:

$$P_t(x,y) = -\sum_{kx} iR_{kx} \exp(-ik_x x) = f_t(x,y) \otimes P/x = g_t(x,y),$$

where $\otimes$ stands for convolution, and P/x is the main value of the generalized function 1/x (the function is not defined for x=0) [8].

Since the contributions of the harmonic processes to the generalized phase portrait have spectral weights R_{kx}, each point (f,g) of the generalized trajectory is formed with weight w(f,g) which is the frequency of occurrence of f and g on the phase plane. One may readily see that w is invariant to the translation-and-scale transformation of the signal $f(x) \dashrightarrow f(mx+b)$ where m and b are transformation parameters. This invariance is due to the fact that the Hilbert image g(mx+b) corresponds to the signal f(mx+b), and w is independent of the coordinate x. Let us form a complex signal s=f+ig. The spectrum of spatial frequencies $R_{kx}(s)$ is single-band one and for $k_x=0$ coincides with the spectrum $R_{kx}(f)$ of signal f. The spectrum of real signal f meets the relations: $\mathrm{Re}R_{-kx} = \mathrm{Re}R_{kx}$ and $\mathrm{Im}R_{-kx} = -\mathrm{Im}R_{kx}$. The signals f and s, thus, have identical degrees of freedom and carry the same quantitative information. Introduction of the signal s, however, allows oné also to introduce qualitative information on the relations between individual components or samples of the signal f due to imposition a causal relation on its samples. As the result, the signals f and s become reciprocally corresponding, and f and the two-dimensional histogram w(s)=w(f,g) are related by a single-values correspondence, i.e. different description w will correspond to different signals f to within the translation-and-scale signal transformations. It should be noted, for example, that a simple histogram w(f) of the signal f does not give a single-valued description, all the samples of f may be "reshuffled" without changing the histogram because w(f) diregards the causal relations between individual signal samples.

At the detection of electromagnetic signals one usually measures not f and g but their non-linear functions F=F(f) and G=G(g). These functions are usually monotone, and, therefore, the passage from the ideal phase portrait w(f,g) to the real one w(F,G) introduces no ambiguity. More difficult is the passage from the line signal description to that of the image as a whole. If the line signals are assumed to be independent, the following description is possible $w_t = \sum_y w(F(y,t),G(y,t))$ which is single-valued to within image line permutations. Description ambiguity may be avoided by merging two description of rows and columns, respectively. The artificial object images are distinguished by the fact that F(f) assumes binary values. In this case the image is describable by a one-dimensional distribution $w_t = \sum_F \sum_y w(F(y,t),G(y,t))$.

The time dependence of phase portrait w is indicative of the stochastic nature of description which is due to the indeterminacy of signal f that may be defined by means of the function $v(f_t)$ of the probability of value f occurrence at time t. Taking into consideration that there is a single-valued relation between the description w_t and signal f_t, one can assume that $v(f_t) = v(w_t)$. In practice, however, the stochasticity of f depends on both the objects in question and the sources of electromagnetic waves. To eliminate the indeterminacy due to the stochastic nature of sources, one has in each particular case to choose suitable functions F and G at the stage of signal detection. With correctly chosen detection techniques, one may assume that function $v_l(w_t)$ describes the frequency of t-th realization of an object belonging to l-th class. The integral description, i.e. a description of l-th class image over the complete observation period, is, therefore, defined by: $W_l(F,G) = \sum_t v_l(w_t)\, w_t$ that defines the conditional density of distribution of F and G of l-th class image and can be used for the Bayes classification of images. The variables F and G themselves can be regarded as informative image attributes or features.

Characteristics of hardware implementing the method.

In the radio-frequency band, the detectors directly record both amplitude, $a_t(x,y)$, and phase, $\Theta_t(x,y)$ envelops of the electromagnetic signals $E(x,y,t) = f_t(x,y) \cos(\omega t) = a_t(x,y) \cos(\Theta_t(x,y) - \omega t)$. Therefore, by passing electromagnetic signal E through a linear filters with pulse characteristics $h_1 = \Delta(x)$ (where Δ - is Diracs function) and $h_2 = P/x$ $F = f = a\cos(\Theta)$, and $G = f \otimes h_2 = g$ may be generated concurrently with F. It is expedient to design transversal filters h_2 for the radio-frequency band with charge-coupled devices and surface acoustic waves that enable filtration within the 10 through 100 MHz band with the dynamic range up to 40 dB [9,10]. These devices seem to be most promising in the communication systems requiring high speeds of image analysis.

In the optical and X-ray bands, when using the most common non-coherent sources of signals and homodyne detection, the signal phase information is lost. As the result, $F = a^2$, and G may be generated only as $G = a^2 \otimes h_2$ or $G = (a \otimes h_2)^2$. This filtration can be implemented with analogue transversal filters (charge-coupled or surface acoustic waves devices) only for robot engineering where high speed of image recognition is a must. Digital methods relying on the FFT hardware, however, offer more promise for such applications as computer-aided design, process control, research, and medicine. They enable filtration in the band up to 10 MHz over 8-bit or 16-bit video signals with special-purpose IC's [11].

In the optical and X-ray bands, when using the coherent sources of signals, the signal phase information including spatial phases of the envelop of the electromagnetic waves is preserved. Therefore, variables $F = f$ and $G = g$ may be generated through heterodyne detection. Hardware filtration $g = f \otimes h_2$ is performed by means of special phase masks or diffraction gratings [12]. These methods offer promise for studies of micro-objects and molecular structures in medicine, biology and materials science.

In practical applications the above methods can complement each other. Of great interest is the fact that linear filtration and histogram analysis may be used for image description. These operations were regarded as effective facilities for image improvement and segmentation, but with the described above method they becme a powerful and practically universal tool of image processing.

Conclusion.

The results obtained in this study have three aspects. The first is a purely technical one. The phase portrait invariance to observation conditions together with the existing techniques of signal normalization allow one to use the description in systems designed for monitoring object state through its own or reflected electromagnetic fields. In doing so, the description weakly depends on the nature of field because it is related to its general physical properties. This generality enables one to allow for the relations between the individual components of the images being analyzed and also reveals the mechanisms of description degeneration at compression of information contained in the original signal that forms the image. Practical generation of the description involves detection and filtration of electromagnetic signals that have been investigated in detail. The algorithmic simplicity of the proposed method and wide variety of technical facilities for its implementation in various applications involving image processing give rise to hopes that it could be used for designing a whole family of devices for analysis of spatial properties of objects through their images.

The second aspect concerns investigation of the visual analyzers of living organisms. In particular, the interaction mechanism of central and peripheral visions so far is not known as well as the role played in it by spatial filtration of visual signals by means of lateral neuron connections. It is, therefore, of interest to compare the description as discussed in this presentation with the model of visual analyzer. For example, the variables F and G may be related to the attributes generated, respectively, by central and lateral visions. Determination of the distribution w(F,G) may be regarded from the standpoint of interaction between these two visions. This hypothesis may by substatiated by the guess that during its evolution the visual analyzer could have completely adapted to the physical properties of the optical band.

The third aspect concerns investigation of complex molecular structures. The above description was obtained measuring the mutual spectrum w(f,g) of the variables f and g having the following interesting features: causal relations between the individual components of the signal $s=f+ig$ under statistical independence of f and g ($\overline{fg} = \sum fg = 0$); the energies of f and g are equal ($\sum f^2 = \sum g^2$). Thus, the description w(s) characterizes the properties of the electromagnetic field $E \cong s$ that are not related to its energy E^*E. This description might prove useful for testing structural changes in complex molecules that do not involve rearrangement of their energy (rotation and vibration) spectra.

Literature

1. M. Hu: Visual pattern recognition by moment invariants. IRE Trans. on Inform. Theor. 8 (1962), 179-187.
2. K. Tanaka, and K. Ozawa: A new type of feature extraction of patterns using coherent optical systems. Pattern Recognition. Pergamon Press. 4 (1972), 251-262.
3. V. Omel'chenko: Foundations of the spectral signal recognition theory. Vyshcha Shkola Publ, Kharkov 1983 (in Russian).

4. A. Rosenfeld: Picture processing. Computer graphics and image processing. 7 (1978) 2, 210-216, 229-235.
5. E. Madelung: Die mathematischen hilfsmittel des physikers. Springer-Verlag, Berlin 1957.
6. P. Dirac: The principles of quantum mechanics. At the Clarendon Press, Oxford 1958.
7. A. Andronov, et al.: A qualitative theory of dynamic systems. Nauka Publ., Moscow 1966 (in Russian).
8. E. Titchmarsh: The theory of functions. Oxford 1932.
9. R. Baertsh, et al.: The design and operation of practical charge-transfer filters. IEEE Trans. ED-23 (1976) 2, 133-141.
10. V. Dmitriev, et al.: Integral piezoelectric signal filters and processors. Radio i Svyaz Publ., Moscow 1985 (in Russian).
11. Inform. of Zorn Corp. firm: A new class of DSP chip - the vector signal processor. Electronics. A Mc Graw-Hill Publ. 59 (1986) 27, 59-66.
12. L. Soroco: Hilbert optics. Nauka Publ., Moscow 1981 (in Russian).

Gy.Cs. Hegedüs: Fast geometric correction of digital images

1. Introduction

Computationally demanding geometric corrections of pictures are often used in digital image processing. In the widest sense they are transformations that from one -or more,geometrically corresponding- INPUT picture(s) generate OUTPUT picture(s) such that the loss of the information of the picture is less than a predefined error limit and simultaneously the distorsion of geometric relationships is less than an another predefined error limit.

The geometric correction can be given in the following two ways.

Theoretically:

If one knows the distorsion of the picture recording system and the relations describing the displacement of the objects recorded or one has some other theoretical knowledge, then one can describe the geometric relation by analytic formulae. In this case the position of each point of the pictures can be corrected exactly.

Experimentally:

The geometric correction can also be defined by points whose INPUT positions as well as their OUTPUT positions are known. These are called reference points or Fiducial marks, and their positions can be experimentally measured. This is not an exact way of generating geometric corrections,as,due to measuring precision, the exact positions of the reference points are not known. The positions of all other points are not known at all, but must be interpolated or extrapolated from corrections found at the reference points.

However the theoretical and experimental approaches can be combined, i.e.,the correction obtained theoretically can be parametrized by calculating the positions of a few reference points.

In section 2, concerning the computational problems, we suggest an efficient solution of the general geometric correction problem. In sections 3 and 4 we discuss how to obtain the correction theoretically and experimentally. Section 5 shows the results of our implementation. A brief bibliography is presented in section 6.

2. Discussion of computational problems

a) Without loss of generality we may assume that if a correction is restricted to an appropriately small domain it may be considered as a linear function e.g. in the case when the distorsion to be corrected is not too large. Linear and affine correction of digital pictures are well developed disciplines. There are several fast algorithms published [2,3] which are based on the fundamental rules of digital straight line representation [1]. It is useful to use these fast algorithms, with appropriate adjusment even in the case of globally nonlinear corrections.

One might use a point-by-point modification but with a smooth correction function one may adjust at much less frequent materials. Theoretically a good solution is to adjust less frequently where actual correction changes slowly and more frequently at other places. Although from the point of view of the number of adjustments needed it is only suboptimal, in our case the following method seems to be promising:

-Let us consider the picture domain e.g. in system A. We can cover the picture with a regular square lattice such that in each square the correction may be considered to be linear. A single square in this lattice will be referred to as a lattice element. The more linear the correction is, the more points of the picture can be included in a single lattice element.

-Obtain the position of each vertex of each lattice element in system B. These are called control data.

-Knowing the positions of the vertices in system B, one can perform the correction as a four-point interpolation at each lattice element of the picture to be correct.

If the correcting relation is invertable, then the transformed lattice elements are disjoint and form a perfect tiling in system B.

This method can be justified as follows:

-To control an adjustment whose density changes, one has to perform extra computations i.e. gradient computation. In the case of a regular lattice it can be omitted.

-Although in this method one has to obtain the control data more often than it would be necessary, since its frequency is determined by the intervals with the fastest change, the amount of extra computation is not considerable since the number of control points is generally much less than the total number of points in the picture.

-If one handles the lattice elements of a row in the system A, then one can raster scan the image. This is compatible with the data structures of many of the common secondary storage systems.

b) In geometric transformations one obtains the position of the vertices of regular lattice defined in one system in another system. If the regular lattice is defined in the INPUT system, then one obtains direct transformation. If it is defined in the OUTPUT system, the one obtains an inverse transformation.

If the correcting relation is invertable, direct and inverse methods will theoretically yield the same result.

From several points of view [4] it is more useful to choose the inverse method, for example in this case every position of the picture in the corrected system is defined i.e. there is no need for initialization.

c) To assign the original and the corrected lattice to each other one needs an interpolating function R. If a spot is homogeneous in the picture, one cannot see the difference between them. However in the case of a picture with outlines there occur considerable differences. The nearest neighbour technique, which is often used in fast algorithms, cannot be used here since it would produce a false outline. More precise results can be obtained by bilinear or spline interpolating functions. However these need much more computation.

It is fruitful to let the picture adjust the choice of the interpolating function. However, this requires the fast detection of changes in the picture. One can detect edges using fixed point arithmetic by convolving the picture function with a quickly decaying function that stresses the local change in the picture.

Thus, when transforming the lattice elements

-one can use a fast algorithm based on a fixed point arithmetic in the homogeneous parts of the picture. In this case the interpolation is the same as in the nearest neighbour technique.

-In addition one can use an additional correction of the position in the less homogeneous parts of the picture, and one should use a more precise interpolating formulae in the neighbourhood of this position.

Thus it is possible to arrange that only the minimal amount of computation is performed to achieve the requested accuracy.

3. A theoretical way of performing geometric corrections

If one knows the correcting relation $f: A \longrightarrow B$, then the steps of correction are as follows.

a) Choose the lattice spacing coefficient according to the smoothness of the correcting relation. It can be done by investigating the extremal values of the gradient function or the coefficient can be determined from prior knowledge.

b) Perform the correction at the vertices of the lattice, and so obtain the control data.

c) Using the control data, perform the row-to-row correction:
-perform the general four-point transformation of the elemental squares using fixed point arithmetic
-process simultaneously the elemental squares including the row which is currently being processed
-use the picture controlled interpolation (incorporating a fast edge detecting technique together with the correction).

4. An experimental way of performing geometric correction

The geometric correction can be defined by obtaining the positions of the points (the so called reference points) which are the same in both the INPUT and in the OUTPUT pictures. Even if corresponding points can be identified unambiguously quantisation of the position limits the accuracy.

In the view of the above points, one should find a correcting relation such that:
-the error of assigning the reference points to one another should be minimal.
-the mapping to be found should be as smooth as possible ever the whole plane of the picture.
-one should be able to obtain a more precise correction by using additional reference points.

Requirements for the algorithm for obtaining the correcting relation are:
-it should require no other information than the positions of the reference points, but if there are such additional pieces of information, then there should be a way to utilize them (such as the parametrization of the correction obtained theoretically by the reference points).
-it should contain both simple and more complex approximations, and one should be able to prescribe the necessary complexity.
-it should converge fast enough even in the case of a large set of reference points.
-it should be possible to control the errors.

Consider a set of reference points that consists of N points and that define a one-to-one correspondence between systems A and B :

$$P_s: \ (X_{sA}, Y_{sA}) \longleftrightarrow (X_{sB}, Y_{sB}) \qquad (s=1,\dots,N) \qquad (1)$$

where P_s denotes the sth reference point, and (X_{sA}, Y_{sA}) and (X_{sB}, Y_{sB}) denotes the coordinates of the positions in systems A and B , respectively. Assuming that measurement error distributions are Gaussian, we must find independent continuous functions

$$f_x: \ \left\{ (X_{sA}, Y_{sA}) \ \middle| \ s=1,\dots,N \right\} \longrightarrow X_{sB} \qquad (2)$$

and

$$f_y: \left\{(X_{sA},Y_{sA}) \mid s=1,\ldots,N\right\} \longrightarrow Y_{sB} \tag{3}$$

such that the errors of correspondence at points P_s $(s=1,\ldots,N)$ are as small as possible and the functions are "smooth" enough. Although the error is not isotropic in this case, the independence yields a considerable decrease in the computation needed.

Assume that function f (f_x or f_y) can be approximated by the sum of m linearly independent, continuous functions $g_i(X_A,Y_A)$ $(i=1,\ldots,m)$:

$$f_a(X_A,Y_A) = \sum_{i=1}^{m} a_i g_i(X_A,Y_A)$$

where f_a is the approximating function of f_x or f_y.

The following holds at the reference points:

$$\bar{f}_a = \sum_{i=1}^{m} a_i \bar{g}_i \tag{4}$$

where $\bar{f}_a$: is a vector with length N. Its entries are the values of variables X_{sB} or Y_{sB}, i.e. they are the values of the approximating function $f_a(X_{sA},Y_{sA})$ at argument (X_{sA},Y_{sA}), $(s=1,\ldots,N)$

a_i : is the ith entry of the coefficient vector with length m. a_i is the coefficient corresponding to function g_i $(i=1,\ldots,m)$

$\bar{g}_i$: is a vector of the values of the function with length N. $g_i(X_{sA},Y_{sA})$ is the value of the ith function at point (X_{sA},Y_{sA}) $(s=1,\ldots,N)$, $(i=1,\ldots,m)$.

The first constraint (accuracy) can be satisfied by minimizing a functional $\Phi(\bar{a})$ that contains the differences in an appropriate way. Thus parameters a_i can be obtained as follows:

$$\min_{a_1,\ldots,a_m} \Phi(\bar{a}) \quad , \quad \bar{a} = (a_1,\ldots,a_m)^T \tag{5}$$

where a common choice is:

$$\Phi(\bar{a}) = \sum_{s=1}^{N} \varphi_s D(\bar{a}) \quad , \quad D(\bar{a}) = (\bar{f} - \sum_{i=1}^{m} a_i \bar{g}_i)^2 \tag{6}$$

$\bar{f}$ denotes the actual value of the approximated function at points P_s $(s=1,\ldots,N)$ φ_s denotes the weights.

The weights φ_s should be choosen according to the distribution of points P_s. φ_s should be smaller where the density of points P_s is greater.

If $m \ll N$, the second constraint (smoothness) can be fairly since, function f_a has relatively few parameters and so when the functions g_i are smooth it has less chance to differ from f.

The smoothness can also be achieved by appropriate regularization. In this case, instead of $\Phi(\bar{a})$, one can use $\Phi'(\bar{a},\xi)$ when obtaining coefficients a_i by minimization:

$$\Phi'(\bar{a}, \xi) = \Phi(\bar{a}) + \xi \cdot \psi(\bar{a}) \tag{7}$$

Functional $\psi(\bar{a})$ must express the smoothness of the function, so it can be, for example, an approximation of the integral

$$\int_K \left| \text{grad } f_a \right|^2 dP \tag{8}$$

where K is the plane of the picture, $p \subset K$.

The choice of parameter ξ determines the balance between whether the approximation is more precise at the points given, or whether the approximating function is smoother.

Unfortunately, both the computation of the functional (computation of the gradient in the whole domain) and the redefinition of it, into a form with maximum gradient using general basis functions, need a lot of computation.

In our case another method seems to be suitable:

a) perform the approximation with $\xi = 0$ i.e. without any constraint of smoothness.

b) try to achieve a smooth approximation by distributing the identification points P_s uniformly and by choosing smooth functions g_i .

c) choose the faster solution rather than optimizing one after performing such global investigations as:
-perform several approximations with different complexities (see later)
-check their smoothness subsequently, by a fast, visual technique, and leave the choices that need trade-offs to human decision.

Note that this method can yield almost the same result as using formula (7), but it is much faster and it can be coupled with the computation of the data directly controlling the correction.

The computation needed by this method depends crucially on m. If one uses approximating function m , one must invert an m by m matrix. Thus, in the case of large m it is difficult to change the basis functions g_i , and to alter the number of functions used in the approximation. Therefore these computations are slow.

This difficulty can be avoided by finding a recursive (but suboptimal) relation between the approximations with m and m+1 terms. Moreover the program should choose the functions used in the m -term approximation from a so called <u>basis function library</u> containing $M \gg m$ functions, such that the error caused by the approximation should be minimal. In this way we do not have to determine the number of terms in an approximation in advance and we use the basis functions that yield the best results at every stage [5].

We also require that the algorithm should utilize the results of the m-term approximation when it computes the m+1 -term, i.e., the magnitude of computation should not increase. Obviously, the number of terms one can use is limited since in the computer a number can be stored with finite accuracy. The algorithm is also required to control the errors.

The recursive algorithm worked out by us yields m approximating functions different from one another and considering the requirements of accuracy and smoothness, we have to select the best one. One gains information about the satisfaction of the requirements of accuracy from the messages of the program implementing the algorithm.

It is useful to couple the check of whether the requirements of smoothness are satisfied with the computation of the control data correction i.e., it is useful that, when performing what is described in Article b) of section 3, the program enables the user to check the result visually. This can be realized by displaying the corrected lattice on a TV-monitor.

The correction can actually be done as it is described in Article c) of section 3.

5. Implementation

In the MIP system (Modular Image Processing, made by SzKI) there is a subsystem for solving the tasks in connections with the geometric corrections.

The approximating method allows us to approximate with a rational fractional function of degree 5, and so the basis function library consists of 21+20 elements. The recursive algorithm uses the elements of a subset of it. Both the subset and the number of the functions used in the algorithm (max. 15) can be specified by the user.

The maximum number of reference points is 255. These points can be defined numerically or interactively by defining a controllable position in the picture displayed.

If the number of reference points is ten then the approximation takes about one minute.

The best approximation is selected by displaying the corrected lattice, producing simultaneously the data directly controlling the correction. Thus the time to investigate an approximation is one minute.

In the case of 256x256 point pictures, the actual correction takes 0.5-2 minutes. This time depends on the need for secondary storage and on the threshold value of the change of interpolation mode.

The time needed is (practically) independent of the complexity of the correction approximated and although the approximation is several magnitudes faster than the precise correction they yield almost identical results.

6. References

[1] B.W.Jordan,W.J.Lemon,B.D.Holm,"An improved Algorithm for the Generation of Nonparametric Curves", IEEE Transaction on Computers, C-22, 1973,1052-1060.

[2] C.Braccini,G.Marino,"Fast Geometrical Manipulations of Digital Pictures", Computer Graphics and Image Processing, 13.,1980,127-141.

[3] Carl F.R.Weiman,"Highly Parallel Digitized Geometric Transformations Without Matrix Multiplications",Proceedings of the 1976. International Conference on Parallel Proceedings, Detroit, Michigan, 1-10.

[4] D.Perny,M.Gangnet,Ph.Coueignoux,"Perspective Mapping of Planar Textures", Computer Graphics,16.,1982.,72-89.

[5] H.L.Ingram,R.Hooker,"The Selection of Approximating Functions for Tabulated Numerical",Technical Report, X-64658,George C.Marshall Space Flight Center, Alabama.

Acknowledgement

The author would like to thank Dr. Géza Álló (Computer Research and Innovation Center,HUNGARY) and Dr. John Illingworth (Rutherford Appleton Laboratory, ENGLAND) for their corrections and useful pieces of advice.

Dr.Gy.Csaba Hegedüs
deputy head of department

Computer Research and Innovation Center
H-1015 Budapest I. , Donáti u. 35-45.
HUNGARY

PARALLEL SIMD CONVOLUTION USING THE FERMAT NUMBER TRANSFORM

Mária Lucká [+], Reiner Creutzburg [++], Hans-Jörg Grundmann [+++], and Marian Vajteršic [+]

1. Introduction

With the rapid advances in large scale integration, a growing number of digital signal processing operations becomes attractive. The convolution property of certain transforms can be used to compute the cyclic convolution of two discrete signals. One such transform is the discrete Fourier transform (DFT) in the modified version of the fast Fourier transform (FFT) /4/ which operates on signals in the complex number field. The FFT of length N requires (N/2) $\log_2(N/2)$ complex multiplications /1/, which take most of the computation time. In addition, the FFT implementation of cyclic convolution introduces significant amounts of round-off error, thus deteriorating the signal-to-noise ratio.
The number-theoretic transform (NTT) was introduced as a generalization of the DFT over residue class rings of integers in order to implement fast cyclic convolution and correlation without round-off errors and with better efficiency than the FFT /2, 3/. Other interesting applications of the NTT are in fast digital filtering /1/, image processing /15/, fast coding and decoding of error-correcting codes /16/, and very fast FFT computation /17/. A large number of transform methods were developed /1, 6/. However, there is one interesting case of the NTT, the Fermat-number transform (FNT), that is well-suited for implementation on a digital computer /2, 5/. In a recent paper /5/ the implementation of the FNT on a sequential computer was described. The aim of this paper is to present a parallel version of the FNT that allows fast computation of cyclic convolutions and correlations on an associative SIMD type parallel computer /11/.
In the second chapter we give a short introduction to the theory of the NTT. The Fermat number transform is studied in detail in chapter 3. The structure of the SIMD type parallel computer and the special **binary arithmetics** are explained in chapter 4. Then, in chapter 5 the parallel associative direct and inverse FNT algorithms are presented. Finally, in chapter 6 the application of the Fermat number transform in convolution computing is shown.

[+] Institute of Technical Cybernetics, Slovak Academy of Sciences, Dúbravská 9, 842 37 Bratislava, Czechoslovakia

[++] Academy of Sciences of the GDR, Central Institute of Cybernetics and Information Processes, P.O.Box 1298, DDR - 1086 Berlin

[+++] Academy of Sciences of the GDR, Institute for Cosmic Research, DDR - 1080 Berlin

2. Number-theoretic transforms

Let Z denote the ring of integers and $m > 1$ an odd integer with prime factorization

$$m = p_1^{r_1} \, p_2^{r_2} \, \dots \, p_s^{r_s}. \tag{1}$$

Then $\alpha \in Z$ is called primitive N-th root of unity modulo m if /7/

$$\alpha^N \equiv 1 \bmod m, \tag{2}$$

$$\gcd(\alpha^n - 1, m) = 1 \text{ for every } n = 1,\dots,N-1. \tag{3}$$

A necessary and sufficient condition for the existence of such primitive N-th roots of unity modulo m is /3/

$$N \mid \gcd(p_1 - 1, \dots, p_s - 1). \tag{4}$$

Note that (3) is always fulfiled if the modulus m is a prime number.

The NTT of length N with α as primitive N-th root of unity modulo m and its inverse (INTT) are defined between N-point integer signals

$$X(n) \equiv \sum_{k=0}^{N-1} x(k)\alpha^{nk} \bmod m, \qquad (n = 0,1,\dots,N-1)$$
$$x(k) \equiv N' \sum_{n=0}^{N-1} X(n)\alpha^{-nk} \bmod m, \qquad (k = 0,1,\dots,N-1) \tag{5}$$

where $NN' \equiv 1 \bmod m$. Note that the components of a signal

$$\underline{x} = (x_0, x_1, \dots, x_{N-1})^T \tag{6}$$

have to be quantized to integers before using the NTT. However in many practical applications this is already done by sensors with analog/digital conversion.

The NTT has a similar structure and properties like the DFT, particularly the cyclic convolution property /3,1/

$$\underline{y} = \underline{x} * \underline{h} \equiv \text{INTT} \left[(\text{NTT } \underline{x}) \circ (\text{NTT } \underline{h}) \right] \bmod m, \tag{7}$$

where $\underline{x}$, $\underline{h}$, $\underline{y}$ are periodic signals of length N of the form (6). Therefore the convolution " $*$ " of two signals $\underline{x}$ and $\underline{h}$ can be performed by taking the inverse transform of the pointwise product " $\circ$ " of the transformed signals.

Consider the well-known cyclic convolution formulae

$$y(n) = \sum_{k=0}^{N-1} h(n-k)\, x(k), \qquad n = 0,1,\dots,N-1. \tag{8}$$

In order to obtain the same result in (7) and (8) it is necessary to choose a unique representation of the m elements of the residue class ring of integers modulo m. Often the symmetric representation of the m elements

$$-\frac{m-1}{2}, \ldots, -2, -1, 0, 1, 2, \ldots, \frac{m-1}{2}, \tag{9}$$

if m is an odd integer, is used. Note that m has to be an odd integer in order to get nontrivial transform lengths N in (4).

If

$$\max |y(n)| \leq \frac{m-1}{2}, \tag{10}$$

then the modulus m is large enough and no overflow modulo m in (7) occurs. Then (9) is called the dynamic range of the NTT. To obtain (10) from (8) follows

$$|y(n)| \leq \max|y(n)| \leq N \max |x(n)| \max|h(n)| \leq \frac{m-1}{2}. \tag{11}$$

If $\max|x(n)| \leq A$, $\max|h(n)| \leq A$, then (11) yields

$$|A| = \sqrt{\frac{m-1}{2N}}. \tag{12}$$

With (12) a simple bound for the signals $\underline{x}$ and $\underline{h}$ is given to prevent overflow in convolution and correlation computations. Better bounds can be obtained using the ℓ_p-norm of the signal $\underline{x}$ /1/. The convolution (7) can be interpreted as a filtering of the signal $\underline{x}$ with the filter $\underline{h}$. For such filtering applications often the following bound is useful

$$|y(n)| \leq \max|x(n)| \sum_{n=0}^{N-1}|h(n)| \leq \frac{m-1}{2},$$

because in most cases the filter $\underline{h}$ and the maximal amplitude $\max|x(n)|$ of the signal $\underline{x}$ are known.

From the numerical point of view one wishes to choose a NTT with parameters N, α, m, that have the following properties:

- the transform length N has to be large enough and highly factorizable (preferably a power of 2),
- the primitive N-th root of unity α modulo m has a simple binary arithmetic,
- the modulus m has to be large enough to avoid overflow and should have a simple binary arithmetic.

For most practical applications α and N are given and a convenient modulus m has to be found, In /6,7/ a useful way is shown to solve this problem by studying cyclotomic polynomials $\Phi_N(x)$ for integer values α and primitive factors of the term $\alpha^N - 1$. The following theorem is given here without proof. For a detailed describtion the reader is referred to /7/.

<u>Theorem:</u> Let $\alpha \in Z$ $(|\alpha| \geq 2)$ and $N > 1$ be given. Further let $m > 1$ be an odd integer. The following conditions are equivalent

a) α is a primitive N-th root of unity modulo m,

b) $\Phi_N(\alpha) \equiv 0 \bmod m, \quad \gcd(N,m) = 1,$ (13)

c) m is a primitive factor of α^N-1. (14)

Other equivalent conditions can be found also in /6/. In (14) $\Phi_N(\alpha)$ denotes the value of the cyclotomic polynomial for $x = \alpha$. A primitive factor of $\alpha^N - 1$ in (14) is an integer m with the following properties

$$m \mid \alpha^N - 1, \qquad \gcd(\alpha^n - 1, m) = 1 \quad \text{for all} \quad 1 \leq n \leq N-1, \tag{15}$$

where gcd denotes the greatest common divisor and | serves for a division operation.

Primitive factors for different α and N are listed in number-theoretic tables /14/. For the definition and properties of the cyclotomic polynomials see /8/. Here we give well-known properties of $\Phi_N(x)$

$$\Phi_p(x) = \frac{x^p - 1}{x - 1} \qquad (\text{ p prime}), \tag{16}$$

$$\Phi_{p^{d+1}}(x) = \frac{x^{p^{d+1}} - 1}{x^{p^d} - 1} \qquad (\text{ p prime, } d \geq 0). \tag{17}$$

In the case $x = 2$ one obtains from (16) the parameters of the Mersenne number transform with modulus $m = 2^p - 1$. For $x = 2$ and $p = 2$ the rule (17) yields the parameters $\alpha = 2$, $N = 2^{d+1}$ for the Fermat number transform with modulus

$$m = 2^{2^d} + 1.$$

In this way the parameters for a lot of other number theoretic transforms can be found. Note that this method can be generalized to arbitrary finite commutative rings with unity /6/.

3. Fermat number transform (FNT)

As stated above, the FNT uses the transform length $N = 2^{d+1}$, the simple binary element $\alpha = 2$ as a primitive N-th root of unity modulo m and the Fermat number modulus

$$m = F_d = 2^{2^d} + 1 = 2^b + 1, \quad (d \geq 0, \ b = 2^d).$$

In table 1 the parameters of the FNT are listed for various d. Also the bound A from (12) is given.

The one-dimensional FNT and its inverse with $\alpha = 2$ are defined as

$$\left.\begin{aligned} X(k) &\equiv \sum_{n=0}^{N-1} x(n)\, 2^{nk} \bmod F_d \\ x(n) &\equiv N^{-1} \sum_{k=0}^{N-1} X(k)\, 2^{-nk} \bmod F_d \end{aligned}\right\} \begin{aligned} &\text{with } N = 2^{d+1}, \ \alpha = 2, \\ &N^{-1} = -2^{2^d-d-1}, \\ &(n,k = 0,1,\ldots,N-1). \end{aligned} \tag{18}$$

The integer $b = 2^d$ denotes the required computer wordlength to represent the m integers in binary form. From table 1 it is obvious

Table 1 Parameters for the Fermat number transform

d	b	$m = F_d$	N (for $\alpha=2$)	A (for $\alpha=2$)
0	1	2^1+1	2	0
1	2	2^2+1	4	0
2	4	2^4+1	8	1
3	8	2^8+1	16	2
4	16	$2^{16}+1$	32	32
5	32	$2^{32}+1$	64	5792
6	64	$2^{64}+1$	128	$2,6\times10^8$
7	128	$2^{128}+1$	256	$8,1\times10^{17}$
8	256	$2^{256}+1$	512	10^{37}
9	512	$2^{512}+1$	1024	10^{75}
10	1024	$2^{1024}+1$	2024	10^{100}

to see that the transform length N is proportional to the wordlength b. This is a great disadvantage of the FNT, if a sequential computer is chosen. For example, a 32-bit machine allows the implementation of a FNT with transform length $N = 128$ and $\alpha = 2$. However, it is possible to extend the transform length by using multidimensional transform techniques /2/, but they are more complex. The associative SIMD type parallel computer /13/ chooses a word length of about 256 bit and is therefore well-suited for the performance of the FNT with longer transform lengths up to 1024.

The following example shows the application of the discrete Fourier- and the Fermat number transforms to one-dimensional cyclic convolution.

Example /5/. Calculation of cyclic convolution of the signals $\underline{x} = (2,1,0,1)^T$ and $\underline{h} = (1,-2,0,0)^T$ of length $N = 4$ via discrete Fourier- and Fermat number transforms, respectively:

a) Discrete Fourier transform. The Fouriermatrix and its inverse are

$$\underline{F} = \begin{bmatrix} 1 & 1 & 1 & 1 \\ 1 & -j & -1 & j \\ 1 & -1 & 1 & -1 \\ 1 & j & -1 & -j \end{bmatrix}, \quad \underline{F}^{-1} = \frac{1}{4}\begin{bmatrix} 1 & 1 & 1 & 1 \\ 1 & j & -1 & -j \\ 1 & -1 & 1 & -1 \\ 1 & -j & -1 & j \end{bmatrix}.$$

This gives the Fourier transforms of $\underline{x}$ and $\underline{h}$

$$\underline{X} = \underline{F}\,\underline{x} = (4,2,0,2)^T,$$

$$\underline{H} = \underline{F}\,\underline{h} = (-1,1+2j,3,1-2j)^T.$$

Pointwise multiplication and inverse Fourier transform gives

$\underline{X} \circ \underline{H} = (-4, 2+4j, 0, 2-4j)^T,$

$\underline{y} = \underline{x} * \underline{h} = \underline{F}^{-1}(\underline{X} \circ \underline{H}) = (0, -3, -2, 1)^T.$

b) Fermat number transform. From (17) and with modulus $F_2 = 2^{2^2} + 1 = 17$ one gets the transform matrix $\underline{W}$ and its inverse $\underline{W}^{-1}$

$$\underline{W} = \begin{bmatrix} 1 & 1 & 1 & 1 \\ 1 & 4 & 4^2 & 4^3 \\ 1 & 4^2 & 4^4 & 4^6 \\ 1 & 4^3 & 4^6 & 4^9 \end{bmatrix} \equiv \begin{bmatrix} 1 & 1 & 1 & 1 \\ 1 & 4 & -1 & -4 \\ 1 & -1 & 1 & -1 \\ 1 & -4 & -1 & 4 \end{bmatrix} \mod 17,$$

$$\underline{W}^{-1} = 4^{-1} \begin{bmatrix} 1 & 1 & 1 & 1 \\ 1 & 4^{-1} & 4^{-2} & 4^{-3} \\ 1 & 4^{-2} & 4^{-4} & 4^{-6} \\ 1 & 4^{-3} & 4^{-6} & 4^{-9} \end{bmatrix} \equiv -4 \begin{bmatrix} 1 & 1 & 1 & 1 \\ 1 & -4 & -1 & 4 \\ 1 & -1 & 1 & -1 \\ 1 & 4 & -1 & -4 \end{bmatrix} \mod 17,$$

and the Fermat number transforms of $\underline{x}$ and $\underline{h}$

$\underline{X} = \underline{W}\,\underline{x} \equiv (4, 2, 0, 2)^T \mod 17,$

$\underline{H} = \underline{W}\,\underline{h} \equiv (-1, -7, 3, -8)^T \mod 17.$

Pointwise multiplication

$\underline{X} \circ \underline{H} \equiv (-4, 3, 0, 1)^T \mod 17$

and inverse transform gives the cyclic convolution of $\underline{x}$ and $\underline{h}$

$\underline{y} = \underline{x} * \underline{h} = \underline{W}^{-1}(\underline{X} \circ \underline{H}) \equiv (0, -3, -2, 1)^T.$ ■

The transform length N of the FNT is always a power of 2, therefore the well-known radix-2 FFT-algorithm can be applied to the FNT /1/. The first step in the radix-2 FFT-algorithm fives for the FNT (18)

$$\begin{aligned} X(k) &\equiv \sum_{n=0}^{(N/2)-1} x(2n)\, 2^{2nk} + 2^k \sum_{n=0}^{(N/2)-1} x(2n+1)\, 2^{2nk} \mod F_d \\ X(k + \tfrac{N}{2}) &\equiv \sum_{n=0}^{(N/2)-1} x(2n)\, 2^{2nk} - 2^k \sum_{n=0}^{(N/2)-1} x(2n+1)\, 2^{2nk} \mod F_d . \end{aligned} \tag{19}$$

With the help of N additions and N/2 multiplications it is possible to perform the transform (18) of length $N = 2^{d+1}$ as 2 transforms of length N/2. For $N = 2^{d+1}$ this is d-times possible. This gives a number of $N \log_2 N$ additions and $(N/2)\log_2 N$ multiplications for a fast FNT. But with the help of a special arithmetic for the FNT it is possible to avoid any multiplication by simple shift operations. This arithmetic is described in the next chapter.

4. Parallel computer system SIMD and the special binary arithmetics

The parallel computer system of the SIMD type enables associative data processing in one of its associative modules. Each module consists of the multidimensional access memory (MDA), arithmetic-logical unit (ALU) and permutation network (PN). MDA memory can be seen as a matrix where respectively rows and columns correspond to the memory words and bit slices. To each word of the memory one processor of ALU is assigned which can perform any elementary logical operation over 1-bit memory slice. In ALU, there are also working registers X, Y and mask register M. Performance of any arithmetic-logical operation over two N-component memory fields of width w follows parallely by words and sequential by bit slices. Parallel complexity is then O (w) and compared with the sequential complexity O (N) the gain in efficiency results for $N \gg w$. The role of PN is to permute components of bit slices by the transport between MDA and ALU. For a more detailed information concerning the construction of parallel algorithms for associative computers a reference is given in /12/. Parallel image processing algorithms for this computer are described in /10/. For a formulation of the algorithms developed some notations will be used. A masked parallel execution of operation $\otimes$ on the fields F1, F2 with result in F3 will be denoted as $F3 \leftarrow F1 \otimes F2$ (M) where respectively 1's (0's) in mask register M correspond to those words in FR where the result should (not) be written in. $\overline{F}$ will be a field with elements obrained by negation of values in a field F. F_i, $(F)_i$, $F[i]$ will denote respectively i-th column slice from the right (i=1,2,...w), i-th component from the top (i=1,2,...q), and i-th q-length block (i=1,2,...,s) of a field F, where w is the width, r is the length and s=r/q is the number of blocks of F. $\emptyset_1$ and I_1 will be used for denoting 1-bit fields of the length l containing respectively 0's and 1's in all components. IN_1 is analogical as $\emptyset_1$ except the first element is 1. By $M \leftarrow$ SHFD $M(2^l)$ will be expressed cyclical shift of a field M 2^l positions down. The data permutation needed in the algorithms will be formulated by notation $F \leftarrow F_{2^{l-1},2^l}$ where: F ist the permuted field of length $r=2^k$, $k > 1$, l-1 is the parameter of permutation performed within the blocks of length 2^l according to the prescription $F[j]_s \rightarrow F[j]_t$ for $j=1,2,\ldots,r/2^l$, $l=1,2,\ldots,k$, $t \equiv (s+2^{l-1}) \bmod 2^l$. The expression $F \leftarrow F.2^{FE}$ will be used for multiplication of elements of a field F by powers of 2 where the exponent is a corresponding value in FE, i.e. $(F)_i \leftarrow (F)_i \cdot 2^{(FE)_i}$ i=1,2,...,r where r is the length of F and FE. The masked operation of locating a value A, into those words of a field F where the corresponding bits of mask M are 1's, will be abbreviated by $F \leftarrow$ LOC A(M). Troughout the paper, log N will be used for the binary logarithm of N being a power of 2.

To perform effectively the fast parallel FNT on the PPS SIMD computer /13/ it is advantageous to use a special binary arithmetics. For computer representation of $m=2^n+1$ numbers, m+1 bits are required. To speed-up the computations by using the wordlength of m bits only, the arithmetics proposed in /2/ will be adopted. In this approach, 1 occurs in (n+1)-st position from the right only if the coded number is 0. Since [1] in the (n+1)-st position correspond to the value $2^n \equiv -1 \pmod m$, in the new representation 0 will correspond to -1. Hence, a number A expressed in the classical binary representation (B-code) as A_B will correspond to the number A_D in the new representation (D-code), whereby

$$A_D = (A-1)_B \ . \tag{20}$$

This change in number coding requires modification of all basic arithmetic operations /9/. We restrict ourselves to show here only the multiplication which is of special importance in the convolution procedure. According to the coding formula (20), the multiplication of two numbers A_D and C_D will be processed as

$$(AC)_D = (AC-1)_B = (A.C)_B - 1, \tag{21}$$

i.e. by multiplying the numbers A,C in the B-code and by final transforming the result into the D-code /9/. A detailed formulation of this and other operations on numbers in D-code from the point of view of the implementation on the PPS SIMD computer will be reported in /11/.

5. Parallel associative direct and inverse FNT algorithm

In this chapter the description of the direct and inverse Fermat Number Transform on the associative type parallel computer will be given. For parallelization of this method the use will be made of the Cooley-Tukey scheme /1/ which has been originated for the Fast Fourier Transform computations /4/. This scheme requires the input data in natural ordering while the output vector is obtained in a bit-reversal ordering. The algorithmical pattern of both the FFT and FNT is the same, the difference being in multiplication by respectively 2^{nk} and w^{nk} (w-complex) in FNT and FFT. Since the exponents remain in both cases the same, the complex multiplication in FFT is replaced by simple shift operations in FNT.

For FNT evaluation of size N on the PPS SIMD, fields F1 and F2 of width N/2+1 are needed for locating respectively the input as well as output data and intermediate results. For storing the exponents a field FE of width log N-1 will be declared in the MDA memory.

Supporting FNT implemented on the associative module with the MDA block consisting of 256 words, it follows that the maximal size of the transform is 256. Hence, the largest set of exponents required are the numbers 0,1,...,127 ordered in the bit-reversal ordering. For locating such ordered number set a field EX of length 128 and width 7 will be declared in the memory. For each $N=2^m$, $2 \leq m \leq 8$, the corresponding set of exponent field EXN is then obtained from first N/2 elements of the field EX assumming only first log N-1 bits from the left.

The parallel algorithm for the direct FNT of the size N can be formulated in following computational stages:

1. Coding the elements of the input vector F1 from B-code into D-code
2. $k \leftarrow 1$
3. Generating the exponent field Fk:
   a) $i \leftarrow 0$
   b) Create the mask
      $Mk \leftarrow (\emptyset_{N/2^k}, I_{N/2^k}, \emptyset_{256-N/2^{k-1}})$
   c) $Fk \leftarrow LOC \ (EXN)_i \ (Mk)$
   d) $Mk \leftarrow SHFD \ Mk(N/2^{k-1})$
   e) $i \leftarrow i+1$
      if $i < 2^{k-1}$ then go to c) else continue
4. $F1 \leftarrow F1.2^{Fk}$
5. $F2 \leftarrow F1_{N/2^k, N/2^{k-1}}$
6. Create the mask
   $Mk' \leftarrow (\emptyset_{N/2^k}, I_{N/2^k}, \ldots, \emptyset_{N/2^k}, I_{N/2^k})$
   $F1 \leftarrow \widehat{F1} \ (Mk')$
7. $F1 \leftarrow F1 + F2$
8. $k \leftarrow k+1$
   if $k \leq \log N$ then go to 3
9. Decoding the elements of F1 from D-code into B-code.

It can be observed that the multiplication in step 4. can be realized on the PPS SIMD by performing the loop for $i=1,2,\ldots,\log N-1$

$$F1 \leftarrow F1.2^{i-1}(Fk_i) \qquad (22)$$

where $\log N-1$ is the width of the exponent field Fk. In other words, only shift operations with masks represented by bit slices Fk_i, $i=1,2,\ldots,\log N-1$ of Fk are required.

The inverse Fermat Number Transform defined by (18) is also based on the Cooley-Tukey algorithmical construction. The requirement of this scheme is to have the elements of the input vector ordered in the bit reversal manner.

The algorithmical scheme is analogical as for the direct transform. The multiplication by negative exponents can be performed as the multiplication by positive ones. Clearly, it holds for a given l

$$2^l \ (-2^{n-l}) \equiv -2^n \equiv -(-1) \equiv 1 \bmod (2^n+1).$$

Hence,

$$2^{-l} \equiv 1.2^{-l} \equiv -2^{n-l} \bmod (2^{n+1}) \qquad (23)$$

where 2^n+1 is the Fermat number.

From a detailed analysis of the fast inverse FNT it follows that the exponents needed for multiplication on base of the formula (23) are 0,127,126,...,1, if the maximal transform length 256 is considered. For a transform of the length $N, N=2^l$, $2 \leq l \leq 8$ the corresponding coefficient field FXN is obtained from first N/2 elements in FX by taking from their binary representation only first $\log N-1$ bits from the right.

The steps of the algorithm for computing the inverse FNT of the size N can be formulated for k=1,2,...,logN in the following manner:

1. Coding the input vector F1 / it is ordered in the bit reversal order/
2. $k \leftarrow 1,\ j \leftarrow N/2$
3. Generating the exponent field Fk of width logN-1:
   a) $i \leftarrow 0,\ l \leftarrow 1$
   b) Create the mask Mk1
      $Mk1 \leftarrow (IN_{2k}, IN_{2k}, \ldots, IN_{2k})$
   c) $Mk1 \leftarrow$ SHFD $Mk1(2^{k-1})$
   d) $Fk \leftarrow$ LOC $(FXN)_i\ (Mk1)$
   e) $l \leftarrow l+1$
   f) if $l > 2^{k-1}$ then go to step 4. else continue
   g) $Mk1 \leftarrow$ SHFD Mk1 (2^0)
   h) $i \leftarrow i+j$
   i) $Fk \leftarrow$ LOC $(FXN)_i(Mk1)$
   j) go to e)
4. $j \leftarrow j/2$
5. $F1 \leftarrow F1.2^{Fk}$
6. $F2 \leftarrow F1_{2^{k-1}\ ,2^k}$
7. Create the mask
   $Mk1' \leftarrow (I_{2^k}, \emptyset_{2^k}, \ldots, I_{2^k}, \emptyset_{2^k})$
8. $F1 \leftarrow \overline{F1}\ (Mk1')$
9. $F1 \leftarrow F1+F2$
10. $k \leftarrow k+1$
    if $k \leqq \log N$ then go to 3.
11. Decoding the elements of F1 from D-code into the B-code.

We note that the multiplication in step 5. can be performed as shown in (22). In step 3., each computational stage for k=1,2,...,logN delivers into Fk exponents located in the field FXN on positions j.l, j=0,1,..., $2^{k-1} - 1$, $l=N/2^k$.

As a consluding step in the inverse FNT algorithm, one has to divide the result by the value N=2n. Since N^{-1} can be expressed from (18) as $N^{-1} = -2^{n-\log N}$; multiplication by this value can be computed by means of formula (23), using shift and negation operations only.

6. Fermat Number Transforms in convolution computing

As mentioned in the chapter 2, the FNT saves the convolution property. It means that the convolution of two periodic signals $\underline{x}, \underline{h}$ can be computed by

$$\underline{x} * \underline{h} = \text{IFNT}\left[(\text{FNT}\ \underline{x} \circ \text{FNT}\ \underline{h})\right] \qquad (24)$$

where IFNT denotes the inverse Fermat Number Transform.

Hence, to evaluate the convolution one needs to perform 2 direct FNT's, one inverse FNT and the point-wise multiplication of the transformed signals. This multiplication is computed in the D-code, according to the formula (22) given in the previous chapter.

The following table summarizes expected times for parallel evaluation of formula (24) on PPS SIMD in dependence on four values of parameter N. For comparison purposes, the table brings also the times for computing the cyclic convolution via the FFT which has been run on this computer with word-length of 32 bits.

As seen, the effectivity gains in favour of the FNT are transparent for N < 256. Analyzing the time estimations for N=256, it is to expect that the lost of efficiency for large N is caused by the increasing word-length of data needed in FNT.

	Times /in sec./ for cyclic convolution using	
N	FNT	FFT
32	0,0057	0,0268
64	0,0141	0,0376
128	0,0360	0,0559
256	0,09500	0,0811

Finally, it is to note that the times given are obtained by counting the instruction costs in corresponding parallel programs . At the conference time we intend to present actual run times obtained by measuring on the parallel computer.

References

/ 1/ Nussbaumer, H. J.: Fast Fourier Transform and Convolution Algorithms. Berlin, Springer 1981

/ 2/ Agarwal, R. C.; C. S. Burrus: Fast convolution using Fermat number transforms with applications to digital filtering. IEEE Trans. Acoust. Speech Signal Process., ASSP-22, 1974,87-97

/ 3/ Agarwal, R. C.; C. S. Burrus: Number theoretic transforms to implement fast digital convolution. Proc. IEEE, 63, 1975, 550-560

/ 4/ Brigham, E.O.: The Fast Fourier Transform. Englewood Cliffs, N.J., Prentice Hall 1974

/ 5/ Creutzburg, R.; H.-J. Grundmann: Schnelle digitale Faltung mittels Fermattransformation. Elektron. Informationsverarb. Kybernet. 21, 1985, 35-46

/ 6/ Creutzburg, R.; M. Tasche: F-Transformation und Faltung in kommutativen Ringen. Elektron. Informationsverarb. Kybernet. 21, 1985, 129-149

/ 7/ Creutzburg, R.; M. Tasche: Number-theoretic transforms of prescribed length. Math. Comp. 47, 1986, 693-701

/ 8/ Lüneburg, H.: Galoisfelder, Kreisteilungskörper und Schieberegisterfolgen. Mannheim, Bibliograph. Institut 1979

/ 9/ Leibowitz, L. M.: A binary arithmetic for the Fermat number transform. Naval Research Lab., Report 7971, Washington,D.C.,1976

/10/ Lucká, M.; M. Muchová, M. Vajteršic: Image processing algorithms for the parallel processor system of the SIMD type. Proc.

Conf. Systems for knowledge and image processing. Smolenice (Czechoslovakia), Nov. 1986, 157-162

/11/ Lucká, M.; R. Creutzburg, H.-J. Grundmann, M. Vajteršic: Parallel associative fast Fermat number transform algorithm and its application. (submitted for publication)

/12/ Miklośko, J.; R. Klette, M. Vajteršic, I. Vrťo: Fast Algorithms and Their Realization on Specialized Computers, Bratislava, Veda Publ. 1985 (in Slovak)

/13/ Richter, K.: Parallel computer system SIMD. In: Proc. Conf. Artificial Intell. Inform. Control Syst. Robots., I. Plander (Ed.), Amsterdam, North-Holland 1984,309-313

/14/ Brillhart, J.; D. H. Lehmer, J. L. Selfridge, B. Tuckerman, S. S. Wagstaff Jr.: Factorizations of $b^n \pm 1$, b=2,3,5,6,7,10,11, 12 up to High Powers. Contemp. Math. 22, Providence,R.I., Amer. Math. Society 1983

/ 15/ Reed, I. S.; T. K. Truong, Y. S. Kwoh, E. L. Hall: Image processing by transforms over finite field. IEEE Trans. Comput. C-26, 1977, 874-881

/16/ Reed, I. S.; R. A. Scholtz, T. K. Truong, I. R. Welch: The decoding of Reed-Solomon codes using Fermat theoretic transforms and continued fractions. IEEE Trans. Inform. Theory IT-24, 1978, 100-106

/17/ Siu, W. C.; A. G. Constantinides: Very fast discrete Fourier transform, using number theoretic transform. IEE Proceedings G-130, 1983, 201-204

THE DESIGN OF OPTIMAL GAUSSIAN DOLP EDGE DETECTORS

G. Sommer [1)] and G. Meinel [2)]

Detection of edges is one of the main tasks in low level computer vision. The general aim is to get a detector of maximal resolution power and stability versus noise in blurred and noisy images with the additional constraint of minimal execution time. These requests are mutually contradicting.

In the last years several authors have argued on the base of psychphysical experiments [1] or by theoretical analysis [2], that the Laplacian of a Gaussian lowpass filter is optimally in the sense of balancing resolution and stability. Additionally a Gaussian is the only function which under any scaling operation remains its uniqueness in location as well in spatial as in frequency domain. This is important for hierarchic detection of edges [3,4]. A serious problem arises from the high time complexity for execution.

We will argue to execute a Laplician of a Gaussian as a difference of Gaussian lowpass (DOLP) filters with different exend by statistical reasons and by its power to minimize the execution time. The adaption to the degradation properties of the image will result in a shift invariant detector which is adapted in a mean sense to the most steepest edges.

1. Local dynamics within a first order Markov model

Edges are loci of high dynamics in the gray level function of an image. This gray level dynamics can equivalently be described by a gradient or curvature. Suppose an undegraded and unblurred image $\underline{f}$ of dimension $\{N \times N\}$ with a probability density function $\underline{p}_f$. Than in the frame of a spatially 1st order Markov model the joint probability function $p_{f\,f}(g_0, g_\tau; \tau)$ with gray levels $0 \leq g_0, g_\tau \leq G-1$ is the adequate description of $\underline{p}_f$. A good estimate of $\underline{p}_{f\,f}$ is given by the in general nonsymmetric cooccurrence matrix $\underline{C}_\tau$

$$\hat{\underline{p}}_{f\,f} = \underline{C}_\tau = \frac{1}{2}\left(\underline{C}_{(-\xi,0)} + \underline{C}_{(0,-\eta)}\right) . \tag{1}$$

$\tau = (-\xi, -\eta) = (-1,-1)$ shall be a special displacement vector in a causal 4-neighbourhood. The gray levels g_0 and g_τ are highly correlated as indicates the concentration of gray levels near the main diagonal of $\underline{C}_\tau$. Decorrelation of these features can be reached by a principal axis transform, which in the case of a 1st order Markov process is

1) Friedrich-Schiller-University Jena, Department of Technology, DDR-6900 Jena, Ernst-Thälmann-Ring 32

2) Medical Academy Dresden, Department of Nuclear Medicine, DDR-8019 Dresden, Fetscherstr. 74

a simple rotation of $\pi/4$ around the center of gravity of $\hat{\underline{p}}_{f\,f}$.

Than the decorrelated principal axis components of $\hat{\underline{p}}_{f\,f}$ are the mean g_s and the gradient g_d of gray levels at each location $(i,j) \in \{N x N\}$

$$g_s = \frac{1}{\sqrt{2}} \left(2g_0 + \left(g_{(-\xi,0)} + g_{(0,-\eta)} \right) \right)$$

$$g_d = \frac{1}{\sqrt{2}} \left(2g_0 - \left(g_{(-\xi,0)} + g_{(0,-\eta)} \right) \right) .$$

Thus by applying the lowpass filter

$$\underline{h}_M = \frac{1}{2} \left(\square_2 + \square_2^T \right) = \frac{1}{4} \left\{ \begin{bmatrix} 0 & 0 \\ 1 & 1 \end{bmatrix} + \begin{bmatrix} 0 & 1 \\ 0 & 1 \end{bmatrix} \right\} = \frac{1}{4} \begin{bmatrix} 0 & 1 \\ 1 & 2 \end{bmatrix}$$

and the gradient filter

$$\underline{h}_D = \underline{\nabla}_2 + \underline{\nabla}_2^T = \begin{bmatrix} 0 & 0 \\ -1 & 1 \end{bmatrix} + \begin{bmatrix} 0 & -1 \\ 0 & 1 \end{bmatrix} = \begin{bmatrix} 0 & -1 \\ -1 & 2 \end{bmatrix} ,$$

we transform the image $\underline{f}$ into its common factors (with respect to the model) homogeneity $\underline{f}_h$ and dynamics $\underline{f}_d$

$$\underline{f} = \underline{f}_h + \underline{f}_d = \underline{h}_M ** \underline{f} + \underline{h}_D ** \underline{f}. \tag{2}$$

In (2) $**$ indicates convolution in two dimensions.

By only changing the support of the filters to an acausal 4-neighbourhood, the principal axis components indicate dynamics as curvature, which is extracted by the Laplacian $\underline{h}_L$. From (2) follows

$$\underline{f} = \underline{f}_h + \underline{f}_d = \underline{h}_{M'} ** \underline{f} + \underline{h}_L ** \underline{f}. \tag{3}$$

with

$$\underline{h}_L = \underline{\nabla}_3^2 + \underline{\nabla}_3^{2T} = \begin{bmatrix} 0 & -1 & 0 \\ -1 & 4 & -1 \\ 0 & -1 & 0 \end{bmatrix} \tag{4}$$

and

$$\underline{\nabla}_3^2 = \underline{\nabla}_2 ** \underline{\nabla}_2 .$$

Equation (3) implies the principle of unsharp masking as

$$\underline{h}_L = \underline{\delta} - \underline{h}_{M'} = \underline{h}_{dolp} . \tag{5}$$

$\underline{\delta}$ is the identity operator

$$\underline{\delta} = \begin{bmatrix} 0 & 0 & 0 \\ 0 & 1 & 0 \\ 0 & 0 & 0 \end{bmatrix}.$$

This means that under the given constraints of the image model and neighbourhood the Laplacian $\underline{h}_L$ is identically with a minimal DOLP filter. Extending the image model to a higher order Markov process and restricting to acausal neighbourhoods this is also valid. Only the Laplacian gets other representations, as will be shown in the next paragraph.

2. Estimation of the local dynamics

The image $\underline{f}$ in (3) may be the result of any restauration of a real degraded image $\underline{f}_d$ with the point spread function of the imaging system $\underline{h}_{PSF}$ and noise $\underline{n}$:

$$\underline{f}_d = \underline{h}_{PSF} ** \underline{f} + n.$$

Normally edge detection relates to $\underline{f}_d$. Under this conditions unsharp masking with (5) as well as direct calculation of the Laplacian is an ill-posed problem and does not suit to estimate local dynamics. To transform it to a well-posed problem it has been shown [2,5] that the convolution with a Gaussian $\underline{h}_G$

$$\underline{h}_B = \underline{h}_L ** \underline{h}_G \tag{6}$$

$$h_{DOLP} = (\underline{\delta} - \underline{h}_M) ** \underline{h}_G = \underline{h}_{G_1} - \underline{h}_{G_2} \tag{7}$$

is nearly optimally. It transforms the highpass filters $\underline{h}_L$ and $\underline{h}_{dolp}$ to the bandpass filters

$$\underline{h}_B = \underline{h}_{DOLP} , \tag{8}$$

which suppress both homogeneity and noise.

Assuming isotropy we have with $\sigma_2 = \alpha\sigma_1, \alpha > 1$,

$$h_G(r) = \frac{1}{2\pi\sigma_r^2} \exp\left(-\frac{r^2}{2\sigma_r^2}\right) \tag{9}$$

$$h_B(r) = \frac{1}{\sigma_r^2}\left(\frac{r^2}{\sigma_r^2} - 2\right) h_G(r) \tag{10}$$

$$h_{DOLP}(r) = \frac{1}{2\pi\sigma_1^2} \exp\left(-\frac{r^2}{2\sigma_1^2}\right) - \frac{1}{2\pi\sigma_2^2} \exp\left(-\frac{r^2}{2\sigma_2^2}\right). \tag{11}$$

This regularization corresponds to the strategy of template matching, which is optimally for any detection task. It minimizes the integral measure of signal power related to the integral measure of noise power. In the case of white noise the template matching filter is given by the deterministic template of the signal to be detected. This is the curvature of an edge in the case of edge detection. Using a Gaussian template has the additional advantage of minimizing the product of resolution and signal-noise ratio, the uncertainty of the detection task.

3. Adaption of a Gaussion DOLP edge detector

To be shift invariant, an edge detector may only be adapted in a mean sense to a class of edges. The adaption of the parameter standard deviation as

$$\sigma_r = \sigma_{PSF} \tag{12}$$

is a natural one, because the point spread function limits to the lower end the edge steepness. Point spread functions often can be modeled as isotrope Gaussians with the standard deviation σ_{PSF}.

Because of isotropy of the PSF we restrict to demonstrate the one dimensional adaption of the Laplacian(10) to σ_{PSF}. Some striking features in spatial and frequency domain characterize the Laplacian. These are the position of zero crossing x_o^L, the position of negative maximum x_{nm}^L, the frequency of the maximum of the transfer function u_m^L and the frequency u_{nm}^L corresponding x_{nm}^L.

$$x_o^L = \sigma_{PSF}, \quad x_{nm}^L = \sqrt{3}\, x_o^L, \quad u_m^L = \frac{1}{\sqrt{2}\pi} \frac{1}{x_o^L}, \quad u_{nm}^L = \frac{1}{2\sqrt{3}} \frac{1}{x_o^L}. \tag{13}$$

The corresponding features of the DOLP filter are with

$$\beta = \sqrt{\frac{2 \ln \alpha}{\alpha^2 - 1}}$$

$$x_o^D = \beta\sigma_1, \quad x_{nm}^D = \sqrt{3}\, x_o^D, \quad u_m^D = \frac{1}{\sqrt{2}\pi} \frac{\beta^2}{\alpha} \frac{1}{x_o^D}, \quad u_{nm}^D = \frac{1}{2\sqrt{3}} \frac{1}{x_o^D}. \tag{14}$$

From the demand of equal zero crossing positions $x_o^D = x_o^L$ follows $\sigma_1 = \frac{1}{\beta}\sigma_{PSF}$ and $x_{nm}^D = x_{nm}^L$, $u_m^D = \frac{\beta^2}{\alpha} u_m^L$, $u_{nm}^D = u_{nm}^L$. Within the range $1.05 \leq \alpha \leq 2.0$ β varies as $1.04 \leq \beta \leq 1.36$ with the result $\sigma_1 \approx \sigma_{PSF}$ and $u_m^D \approx u_m^L$. Therefore the Gaussian DOLP edge detector is equivalently to the Laplacian of a Gaussian. According Marr and Hildreth [1] a ratio $\alpha \approx 1.6$ balances detector energy and computation time for a Gaussian DOLP filter.

4. Synthesis of a Gaussian DOLP edge detector

To reduce quantization errors for filter coefficients of $\underline{h}_G$ within an odd finite support $\{KxK\}$, the computational burden are K^2 multiplications with $K \approx 8\,\sigma_r$. A general principle of reducing the time complexity of 2D-FIR-filters does not exist, because they are in general irreducible [6]. The Gaussian breaks this, because of the existence of the central limit theorem. This states that repetitive convolution of any root tends to approximate a Gaussian [7]. Therefore by using the identity (8), the Laplacian (6) can be realized as DOLP filter (7) with synthesised $\underline{h}_{G_1}$ and $\underline{h}_{G_2}$ to a certain limit, by using suitable roots.

A general root $\underline{h}_k$ of dimension $\{kxk\}$ yields after n selfcascading steps an approximation of a Gaussian

$$\hat{\underline{h}}_G = \underline{\Gamma}_K = (\underline{h}_k)^{**n} = \underline{h}_k^{(1)} ** \underline{h}_k^{(2)} ** \dots ** \underline{h}_k^{(n)} \tag{15}$$

with $K = n(k-1)+1$ and a gain relating multiplications of $\gamma = \frac{1}{n}\left(\frac{K}{k}\right)^2$.

The most simple root is a 2D boxfilter $\underline{\sqcap}_k$, whose coefficients are identic one. Thus on principle the convolution of $\underline{f}$ with $\underline{\sqcap}_k$ can be realized without any multiplication. We used the recursive boxfilter algorithm given by McDonnell [8] with a complexity of 1 multiplication and 4 additions per pixel independent of k.

The variance of $(\underline{\sqcap}_k)^{**n}$ is approximately

$$\sigma_{kn}^2 = \frac{nk^2}{12} \quad . \tag{16}$$

Therefore the convergence to a given K rises with k^2.
Another interesting root is the triangle filter

$$\underline{\Lambda}_3 = (\underline{\sqcap}_2)^{**2} = \begin{bmatrix} 1 & 1 \\ 1 & 1 \end{bmatrix} ** \begin{bmatrix} 1 & 1 \\ 1 & 1 \end{bmatrix} = \begin{bmatrix} 1 & 2 & 1 \\ 2 & 4 & 2 \\ 1 & 2 & 1 \end{bmatrix} , \tag{17}$$

which needs shifts instead of multiplications.
Remembering the band limitation of the image caused by $\underline{h}_{PSF}$, further convergence to the Gaussian can be reached by application of an expansion operation $E_m\{\underline{h}_k\}$. This corresponds to a sampling function

$$S_m(i,j) = \sum_{q=-K_z}^{K_z} \sum_{r=-K_z}^{K_z} \delta(i+mq, j+mr) \tag{18}$$

with $K = 2K_z+1$. This operation causes a new Nyquist frequency $u_N^m = \frac{1}{m}u_N$, which yields a reflection of the power spectrum at this frequency. This would result in artefacts and amplified noise. But these aliased image components can be compensated by cascading with another expanded filter.

Expansion yields a filter $\underline{h}_{k^+}$ with $k^+ = m(k-1)+1$ and a gain $\gamma = \left(\frac{k^+}{k}\right)^2$.

Under the constraint of band limited signals expansion and cascading are commutatively:

$$E_m\{(\underline{h})^{**n}\} = (E_m\{\underline{h}\})^{**n}. \tag{19}$$

Some simple production rules demonstrate the power of this operation for k = 1,2,3, ...

$$E_k\{\sqcap_k\} ** \sqcap_k = \sqcap_{k^2}$$

$$E_{k+1}\{\Lambda_{2k+1}\} ** \Lambda_{2k+1} = \Lambda_{2(k+1)^2-1}$$

$$(E_{2^k-1}\{\Lambda_3\})^{**n} = \Lambda_{2^{k+1}-1} \;.$$

In figure 1 the impulse response of the DOLP edge detector

$\underline{h}_{DOLP} = \Gamma_9 - \Gamma_{13}$ and those of the lowpass filters

$$\Gamma_9 = (\Lambda_3)^{**2} ** E_2\{\Lambda_3\}$$

$$\Gamma_{13} = (\sqcap_3)^{**2} ** (\sqcap_5)^{**2}$$

are shown. The convolution of $\underline{f}$ with Γ_9 is realized in three passes by using quadrant symmetry of Λ_3 with 9 shifts and 24 additions per pixel. The convolution with Γ_{13} needs four passes with a total of 4 multiplications and 16 additions per pixel.

Figure 2 shows the transfer functions H(u) of those filters and the modulations transfer function MTF of the imaging system, an Anger scintillation camera, which is used in Nuclear Medicine. The upper boundary frequency of the MTF is indicated as u_g. Because $u_g \approx u_N^2$ the influence of aliasing onto true image contrast can be neglected.

5. Design of a Gaussian DOLP edge detector

The design of Gaussian DOLP filters by synthesis on the base of cascaded and expanded boxfilters means the multiplication of transfer functions

$$H_\Gamma(u) = \prod_i H_i(u)$$

with

$$H_i(u) = H_{\sqcap_k}(u) = \frac{\sin(k\pi mu)}{\sin(\pi mu)}$$

under the constraints of minimal sideband energy, minimal cascading steps and minimal computation time. This is a simple variation problem with concentration of the attention to the neighbourhood of precomputed sideband positions. Those are given for n = 1,2,3, ..., n^* within the range $0 < u \leq \frac{1}{2}$ by

$$u_n^{sb} = \frac{n}{(k-1)m} \;.$$

Zeros of the transfer function are at frequencies

$$u_n^z = \frac{n}{km} \;.$$

The number of zeros and sidebands is given by

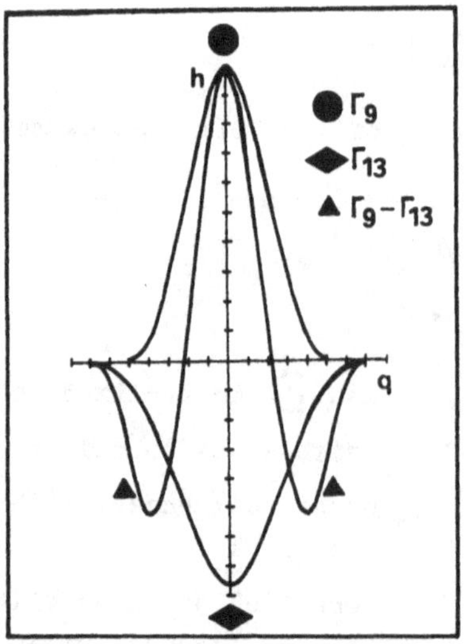

Fig. 1. Impulse responses of a Gaussian DOLP detector and of approximated Gaussian lowpass filters

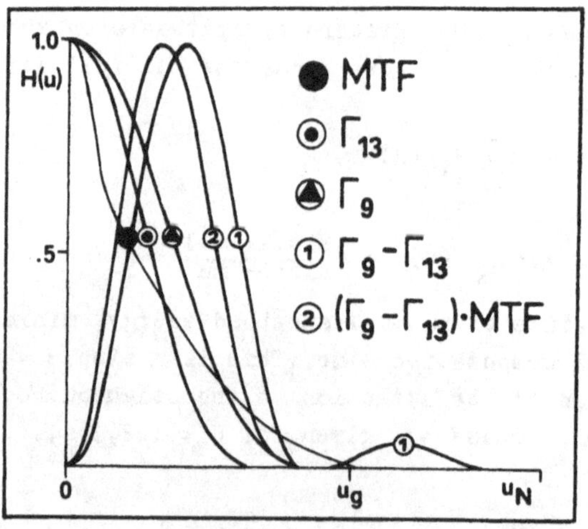

Fig. 2. Transfer functions corresponding fig. 1 and the influence of the MTF onto the output of the edge detector

$$n^* = \frac{m}{2} (k-1).$$

If cascading of root filters has a remainder of some percent sideband energy beyond the boundary frequency like ① in fig. 2, this will contribute with a noise component to the output. Therefore such effects should be avoided.

By precomputation of a bank of filters with different zero crossings the method will be adapted to different types of edges in an image.

6. Literature

1 Marr, D.,E. Hildreth: Theory of Edge Detection. Proc. R. Soc. Lond. B207 (1980) 187-217

2 Torre, V., T.A. Poggio: On Edge Detection. IEEE Trans. Patt. Anal. Mach Intell. PAMI-8 (1986), 2, 147-163

3 Burt, P. J.: Fast Filter Transforms for Image Processing. Comp. Graph. Image Process. 16 (1981) 20-51

4 Crowley, J. L., R.M. Stern: Fast Computation of the Difference of Low-Pass Transform. Techn. Rept. CMU-RI-TR82-18, The Robotics Institute, Carnegie-Mellon Univ., Pittsburg, Nov. 1982

5 Canny, J.F.: Finding Edges and Lines in Images. Techn. Rept. MIT-AI-TR-720, Artif. Intell. Lab., MIT, Cambridge, June 1983

6 Hayes, M.H., J.H. McClellan: Reducible Polynomials in More Than One Variable. Proc. IEEE 70 (1982) 2, 197-198

7 Sommer, G., G. Meinel: Probleme der Implementierung lokaler Laplace-Filter. Wiss. Beiträge IH Dresden, 15 (1985)4, 59-76

8 McDonnell, M.J.: Box Filtering Techniques. Techn. Rept. TR-966, Comp.Sci. Center, Univ. of Maryland, College Park, Nov. 1980

Parallel matrix multiplication on an array-logical processor

R.-J. Vilser ,R. Creutzburg ,H.-J. Grundmann ,M. Goessel

1. Introduction

In this paper we describe two types of blockwise matrix multiplication of a gray scale image P by use of the Displayprocessor (DIP) of the Robotron image processing system A6472 / 1 /. The image P is partioned into subimages Pi of N**2 pixel each. Every subimage Pi is multiplied by the same (NxN) - matrix T of integer coefficients. The generalization to (N1xN2)-matrices , (N1≠N2) , is without differences but not presented here. In the use of right side multiplication the result is an image Q of subimages Qi of N**2 pixels each with

$$Qi = Pi * T .$$

In the case of left side multiplication the result is an image R of subimages Ri with

$$Ri = T * Pi .$$

N is a submultiple of L=512, where L denotes the length of the sides of the considered image P. The presented parallel algorithm for blockwise multiplication of the complete (512x512)- gray scale image needs N**2 or 2N**2 passes through the DIP (each 80 ms) in dependence on the desired accuracy of the result and the number of available memory blocks. In this time 2**27 integer multiplications and 2**27 -2**18 additions are executed. To achieve a high accuracy residue arithmetics is used. The applications till now are transforms (blockwise Walsh transform, Hadamard transform, ...) and image compression.

2. Right side and left side blockwise matrix multiplication

The blockwise matrix multiplication will be explained now. As an example the image P is partitioned in (2x2)-subimages and each of this subimages is multiplied by the given (2x2)-matrix T of integer coefficients. N=2 is chosen here for the simplicity of representation. Then we have

$$P = \left[\begin{array}{cc|cc|c} P11 & P12 & P13 & P14 & \dots \\ P21 & P22 & P23 & P24 & \dots \\ \hline P31 & P32 & P33 & P34 & \dots \\ P41 & P42 & P43 & P44 & \dots \\ \hline \dots & \dots & \dots & \dots & \dots \\ \dots & \dots & \dots & \dots & \dots \end{array}\right] , \quad T = \begin{bmatrix} T11 & T12 \\ T21 & T22 \end{bmatrix}$$

with the blockmatrices

---

1) Academy of Sciences of the GDR, Central Institute of Cybernetics and Information Processes, P.O.Box 1298
DDR - 1086 Berlin

2) Academy of Sciences of the GDR, Institute of Cosmic Research, Satellite Ground Station, Kalkhorstweg,
DDR - 2080 Neustrelitz 5

$$\begin{bmatrix} P11 & P12 \\ P21 & P22 \end{bmatrix} , \begin{bmatrix} P13 & P14 \\ P23 & P24 \end{bmatrix} , \ldots .$$

The result for right side multiplication is

$$\left[\begin{array}{cc|cc|c} P11.T11+P12.T21 & P11.T12+P12.T22 & P13.T11+P14.T21 & P13.T12+P14.T22 & \ldots \\ P21.T11+P22.T21 & P21.T12+P22.T22 & P23.T11+P24.T21 & P23.T12+P24.T22 & \ldots \\ \hline P31.T11+P32.T21 & P31.T12+P32.T22 & P33.T11+P34.T21 & P33.T12+P34.T22 & \ldots \\ P41.T11+P42.T21 & P41.T12+P42.T22 & P43.T11+P44.T21 & P43.T12+P44.T22 & \ldots \\ \hline \ldots \quad \ldots & \ldots \quad \ldots & \ldots \quad \ldots & \ldots \quad \ldots & \\ \ldots \quad \ldots & \ldots \quad \ldots & \ldots \quad \ldots & \ldots \quad \ldots & \end{array}\right]$$

It is possible to see that the columns 0,2,4,... of the image P are multiplied by T11, columns 1,3,5,... by T21, the results are placed in columns 0,2,4,... of the resulting image Q and added. Furtherone columns 0,2,4,... of the image P are multiplied by T12, columns 1,3,5,... are multiplied by T22, the results are placed in columns 1,3,5,... of the resulting image Q and added. In general every N-th column of the image P is multiplied by a constant factor of T, placed in the corresponding column of the resulting image Q and added to the previous results in this column of Q. These operations can be efficiently implemented on the SIMD machine DIP / 1 / by use of an image mask S and look-up tables. In the image mask S every N-th column is marked. (This image mask is generated by the host computer). By a simple logical combination of the image mask and the image P which is to be processed, only the correct selected columns of P will be multiplied by the corresponding coefficients of T. The multiplication of selected columns of P by a constant factor Tij from T is realized with the help of look-up tables and an addition in the corresponding ALU of the DIP channel. In the DIP channels such data are combined which synchronically flow through the pipeline. For every coefficient of the matrix T only one pass through the DIP is necessary.
The image memories can be red with arbitrary read-offset values, and it is possible to write into the image memory with write-offset values. To control which data are combined in the DIP channels only the read-offsets are used here. By this technique it is possible to "shift" the image mask over the image P step by step and to guarantee the correct placement of the results. In our example the image mask is simple a matrix in which every even column contains only 0's and every odd column contains only 1's, in the most significant bit position. The following algorithm shows the blockwise matrix multiplication (right side multiplication) in a FORTRAN-like form, but takes into account that R [x,y] , P [x,y] and S [x,y] are notations of image memories with read offsets (row- and column offsets) as described in / 2 /. This is the algorithm for right side blockwise matrix multiplication:

```
        INTEGER*2 T(N,N)
        R [0,0]  =0
        DO 10 I=0,N-1
        DO 10 J=0,N-1
10      R [0,0]  =R [0,0] +(P [J-I,0] .AND.S [-I,0] )*T(J+1,I+1) .
```

R is the image memory where the result is accumulated, P is the image memory where the image P which is to be processed is stored, S is the image memory for the image mask, T(N,N) is the field of (NxN)-coefficients of the matrix T. For the left side multiplication instead of a column mask S a row mask Z is used. Every N-th row of the mask Z is marked by 1, all the other elements of Z are marked by 0. Again a FORTRAN-like notation is used. This is the algorithm for left side blockwise matrix multiplication:

```
         INTEGER*2 T(N,N)
         R [0,0]  =0
         DO 10 I=0,N-1
         DO 10 J=0,N-1
10       R [0,0]  =R [0,0] +(P [0,J-I] .AND.Z [0,-I] )*T(I+1,J+1) .
```

Of course these are symbolic descriptions of the corresponding programs for the DIP only. The real programming of the DIP is described in / 2 /. In the described algorithm the data pass the channels of the DIP many times. Because it is only possible to read data from an image memory and to write into another image memory, two image memories are needed to accumulate the results. The real state of the channels of the DIP for blockwise matrix multiplication is shown in Fig. 1. The image P is stored in image memory RM2. RM3 contains in its most significant bit plane the column mask S or the row mask Z, resp. The result is accumulated by use of the image memories RM0 and RM1. The look-up tables E1 and E2 are defined by FORTRAN subroutines. By use of E1 the multiplication of the incoming data with the corresponding constant is executed. E2 and the logical AND of ALU2 guaranty that only every N-th column (or row) is selected. The previous result and the new incoming result are added in ALU3. The look-up table A is used if residue arithmetics is used. The word length of the image memories is 8 bits only. Therefore the accuracy of 8 bits of the result of the simple algorithm may be not high enough for some applications. In the next section we show how to achieve an accuracy of 14 or 16 bits by use of residue arithmetics.

3. High accuracy and residue arithmetics

Instead of ordinary multiplication and ordinary addition multiplication and addition is now performed modulo p1 and modulo p2 where p1 and p2 have to be relatively prime integers. In the case of 4 image memories p1=127 and p2=128 are used. x mod 127 and x mod 128 are 7-bit numbers. $\langle x \rangle_p$ denotes x modulo p. The 8-th most significant bit of the memory where $\langle x \rangle$ 127 or $\langle x \rangle$ 128 are stored, is used to represent the mask information (column mask S and row mask Z). In the case of 8 image memories p1=255 and p2=256 are chosen. Therefore for the mask information an additional memory is needed.
We describe here the 4 memory version only. The look-up tables are defined now by the following FORTRAN notation:

```
C
C I,J are the same integers described in the previous algorithm
C
         INTEGER*2 E1(256),E2(256),A(256),M,C
         DO 20 K=0,255
         C=K*T(I+1,J+1)
20       E1(K+1)=C - M*(C/M)
         DO 30 K=0,127
30       E2(K+1)=0
         DO 40 K=128,255
40       E2(K+1)=255
         DO 50 K=0,255
50       A(K+1)=K - M*(K/M) .
```

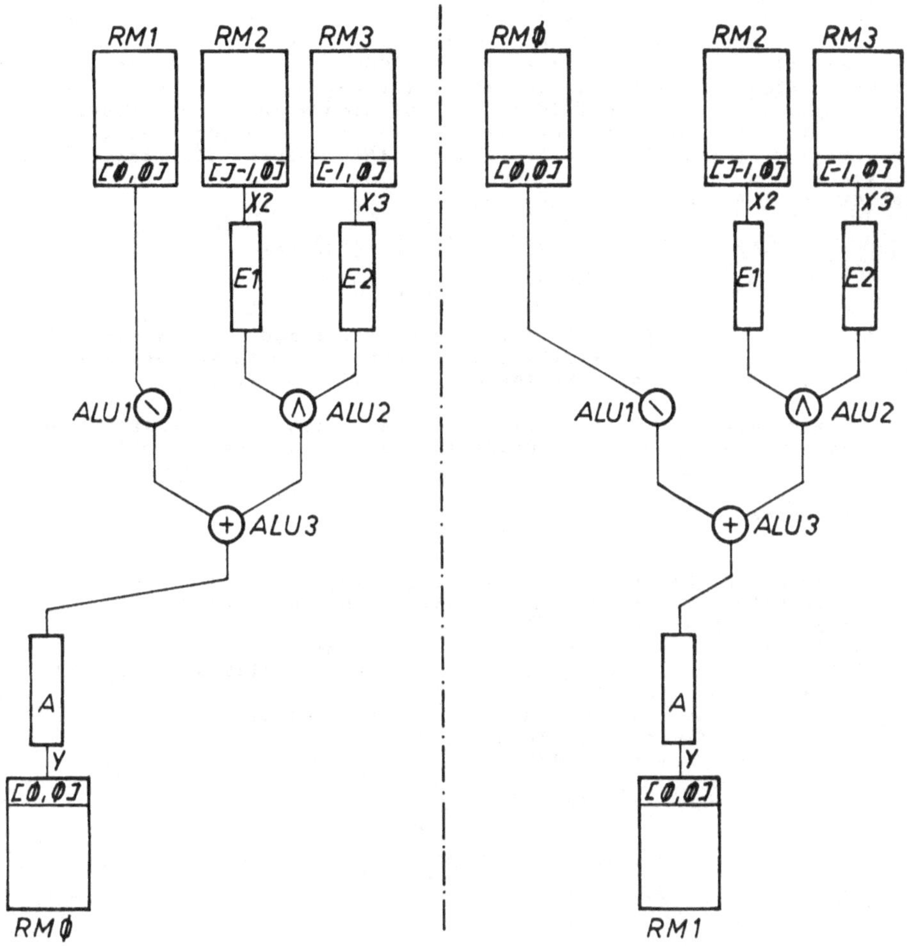

Channel "Red" Channel "Green"

Fig.1 Calculation scheme of blockwise matrix multiplication modulo 127 and modulo 128 on DIP

For p1=127 and p2=128 we have M=127 and M=128, resp. At first the blockwise matrix multiplication modulo p1=127 is computed. The effect of the look-up table ET2 is quite simple. If the most significant bit (i.e. the mask information) equals 1 its output is 255 or in binary representation 11111111 otherwise 00000000. ALU2 is acting as componentwise conjunction. Therefore the output of ALU2 equals to its left input if the mask bit is 1 and 0 otherwise. The modulo 127 result together with the row mask is stored in RM3 and the modulo 128 result is derived. The modulo 127 and the modulo 128 results are converted to a positional number representation by use of the following equations

$$x = \langle x \rangle_{128} + 128 \cdot a1 \quad \text{for } 0 \leqq x \leqq 127 \cdot 128 - 1,\; a1 < 127$$

$$a1 = \langle a1 \rangle_{127} = \langle \langle x \rangle_{127} - \langle x \rangle_{128} \rangle_{127} .$$

which are very similar to the known mixed-radix-conversion /3/ process, but with carefully separated bits. $\langle x \rangle_{128}$ are the least significant 7 bits and a1 is the most signifcant bit.

The determination of the 7 most significant bits can be easily done in one pass of the DIP by an appropriate use of the look-up tables.

4. References

/1/ V. Kempe ;F. Kenner; B. Rebel; W. Schulze and W. Wilhelmi : Interaktives Bildverarbeitungssystem robotron A647x Neue Technik im Buero 26 (1982) No. 5, pp 131-137

/2/ A. Graf: Modelle und Programmierung des A6470. Internal Report, Central Institute of Cybernetics and Information Processes, Berlin 1985

/3/ N.S. Szabo and R.J. Tanaka: Residue Arithmetics and its Application to Computer Technology. McGraw Hill, New York 1967

AFFINE TRANSFORMATIONS ON IMAGES REPRESENTED BY A LIST DATA STRUCTURE

Luth, Nailja[1)]

1. Introduction

The geometrical transformations play an important role in many tasks of digital image processing. Some examples of such tasks are the geometrical correction of images, image calibration, but also the merging of image fragments into a new image.
In this paper we consider the class of affine transformations, which are used especially often.
The general equation of this kind of transformations is given below:

$$\begin{pmatrix} x' \\ y' \end{pmatrix} = \begin{pmatrix} a_1 & a_2 \\ b_1 & b_2 \end{pmatrix} \cdot \begin{pmatrix} x \\ y \end{pmatrix} + \begin{pmatrix} a_3 \\ b_3 \end{pmatrix} \qquad (1.1)$$

We shall consider two ways to realize such transformations (scaling, rotation, translation and reflection).
The geometrical transformations are realized on a Robotron A6472 image processing system with the special SIMD-pipeline processor DIP /1/ . Fast parallel algorithms of image rotation, scaling and reflection have been developed for the DIP, working in integer arithmetics /2/ .
The analysis of the transformation of raster images has shown, that these algorithms are not good enough for use in many cases, e.g. for the transformation of segmented images with small details.
An alternative way is based on the manipulation of segmented images on a high level description. This way is based sufficiently on the investigation of Kovalevsky on digital topology and the description of segmented images as a list structure /3//4/ .

2. Affine transformations of segmented raster images.

The effects of a rotation of a binary image are shown in figures (1-3). Three algorithms of raster trasformations are tested.

1)
Dipl.-Math. Nailja Luth
Academy of Science of the GDR
Central Institute on Cybernetics and Information Processes
Kurstraße 33, Berlin 1086, GDR

Fig. 1
Original image

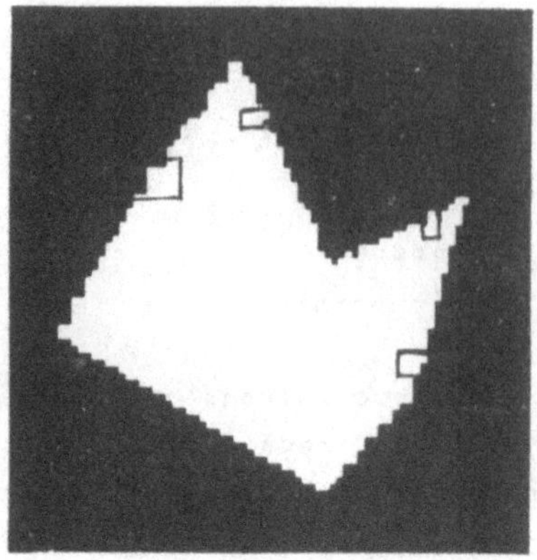

Fig. 2
Image rotated (α =30°)
by 3 shearings

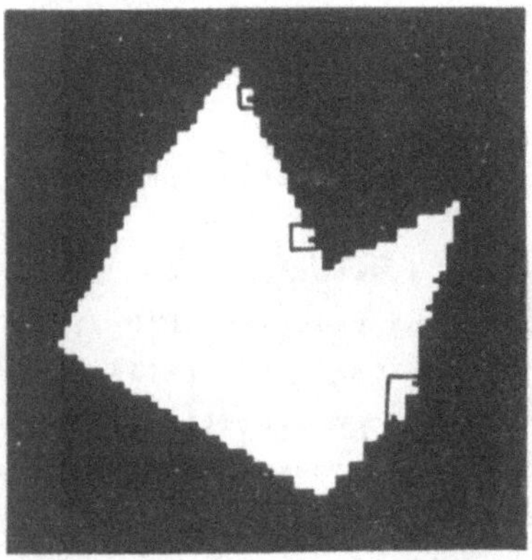

Fig. 3
Image rotated by (α =30°)
stretching, 3 shearings, shrinking

Fig. 4
Image rotated (α =30°)
on the list structure

The first method is based on the representation of the rotation as a superposition of three shearing operators:

$$T = \begin{pmatrix} \cos\alpha & \sin\alpha \\ -\sin\alpha & \cos\alpha \end{pmatrix} = \begin{pmatrix} 1 & a \\ 0 & 1 \end{pmatrix} \cdot \begin{pmatrix} 1 & 0 \\ b & 1 \end{pmatrix} \cdot \begin{pmatrix} 1 & c \\ 0 & 1 \end{pmatrix},$$

where $a=c=(1-\cos\alpha)/\sin\alpha$ and $b=-\sin\alpha$.

The three shearing operations are realized on the A6472 system by an integer approximation (s. Fig. 2). A modification of this method uses previous stretching of the image and the respectively shrinking process after the rotation (s. Fig. 3). In both cases the rounding process can degrade straight lines and destroy the connectivity of narrow regions. The realization of a rotation algorithm in the floating point arithmetics is sufficiently more time expensive. We search for every point of the raster grid the preimage point in the origin. The number of errors decrease, but nevertheless the connectivity of regions with a width of one pixel may be destroyed because of the final roundation process.

All three method do not satisfy the principle of superposition and therefore every origin can be transformed only once.

3.Affine transformations on a data list structure.

The proposed method is sufficiently based on the representation of a segmented image as a list structure. This structure, called "CELIST", contains metrical as well as topological information . The topology of the raster image was proposed by Kovalevsky /3//4/.

The topology contains three kinds of basic elements: 0-dimensional elements (pels) ,1-dimensional elements (lels), 2-dimensional ones (arels). In this topology a segmented raster image is regarded as a so called cellular complex. A connected set R_i of elements with a unique gray level g_i is called a region. Every region R_i represents in the used topology a 2-dimensional cell with its 1-dimensional boundary cell S_i .

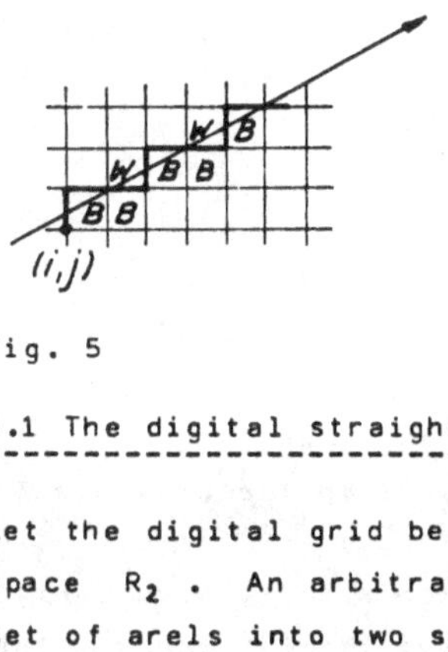

Fig. 5

S_i is a sequence of line elements (lels) and pels. A graphic representation is given in Fig. 5. r_i and r_j denotes the the gray level of the coresponding regions R_i and R_j . The boundaries S_i and S_j are shown as lines.

3.1 The digital straight segment and its Euclidean preimage.

Let the digital grid be embedded into the 2-dimensional Euclidean space R_2 . An arbitrary Euclidean straight segment (ESS) divides the set of arels into two subsets. Moving along the ESS into a predefined direction we label all arels on the right side of the ESS with black color and all arels on the left side with white color. Arels, which are crossed by the ESS, get the color according to the fact, where the most area of the arel lays. The border dividing the grid into a black and a white half gives us a sequence of lels called a digital straight segment (DSS). We say that the DSS is the digital representation of the ESS and the ESS is the preimage of this DSS. The algorithm of constructing the DSS satisfy the condition of minimal distance of the pels from the ESS. (see Fig.6)

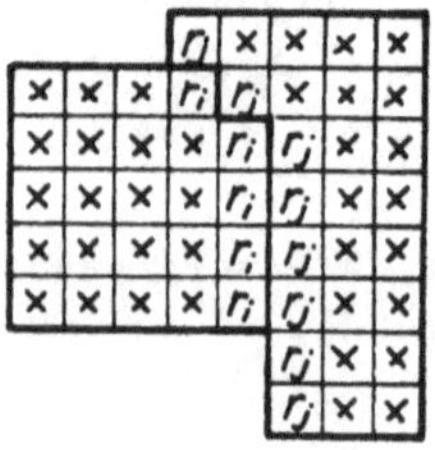

If we have built up the sequence of lels up to pel(i,j) we have to select the next lel from two possible directions according to the procedure mentioned before.

Fig. 6

We can describe this process analytically as follows:

Given an ESS

$$y = y_S + k(x - x_S) \qquad (1)$$

defined by the start point (x_S, y_S) and the end point (x_E, y_E).

$$k = (y_E - y_S)/(x_E - x_S)$$

The starting point of the DSS is defined by

$$X_1 = [x_S + 0.5]$$
$$Y_1 = [y_S + 0.5]$$

The DSS can be represented by a sequence $\{t_n (X_n, Y_n)\}_{n=1}^{N}$ with

$$t_{n+1} = \begin{cases} 1, & \text{for } y(X_n + 0.5) \geqslant Y_n + 0.5 \qquad (2.1) \\ 0, & \text{for } y(X_n + 0.5) < Y_n + 0.5 \qquad (2.2). \end{cases}$$

The coordinates of (X_n, Y_n) are given by

$$X_n = X_1 + \sum_{i=1}^{n} (1 - t_i)$$
$$Y_n = Y_1 + \sum_{i=1}^{n} t_i \; .$$

The condition (2.1) is equivalent to the condition, that the distance from the point(X_n, Y_n) to (X''_{n+1}, Y''_{n+1}) is larger then the distance to (X'_{n+1}, Y'_{n+1}).

We provide the proof for an ESS with $K \leq 1$ and $x_E > x_S$, $y_E > y_S$ only. Let be R_1 the distance from point (X'_{n+1}, Y'_{n+1}) to the ESS and R_2 the distance from the point (X''_{n+1}, Y''_{n+1}) to the ESS.

From Fig.7 we get

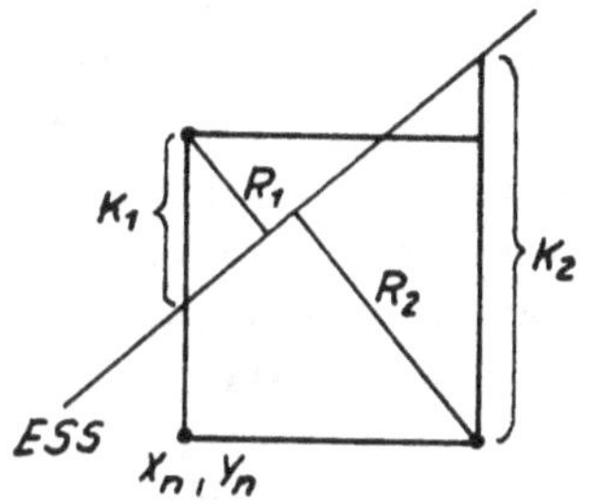

Fig. 7

$$R1 \leq R2 \iff K_1 \leq K_2$$

$$K_1 = Y'_{n+1} - y(X_n)$$

$$K_2 = y(X''_{n+1}) - Y_n \, .$$

We have

$$Y'_{n+1} - y(X_n) \leq y(X''_{n+1}) - Y_n$$

$$y(X''_{n+1}) + y(X_n) \geq Y'_{n+1} + Y_n$$

and because $Y'_{n+1} = Y_n + 1$

$$X''_{n+1} = X_n + 1$$

$$(y(X''_{n+1}) + y(X_n))/2 \geq (2Y_n + 1)/2$$

$$y(X_n + 0.5) \geq Y_n + 0.5$$

q.e.d.

The ESS defines the respective DSS uniquely. On the other hand if we define the preimage ESS of an given DSS in such a way, that the mentioned condition is satisfied the respective DSS' is identically to the previous DSS.

3.2 Manipulations on the list structure.

The sequence of lels S_i is coded by a polygon. The movement along the boundary is fixed as a clockwise, i.e. the inner part of a region lays allways to the right hand side of the boundary. Algorithms for the recognition of digital straight segments have been developed by Kovalevsky and Wieczorek /5/. Every recognized DSS is approximated by its preimage ESS. The corresponding records in the list structure "CELIST" are defined as the crossing points of two adjancent DSS. For every I the boundary S_i is represented in the "CELIST" by its preimage polygon. We can regard the whole cellular complex as a preimage of the segmented raster image I.

This preimage allows us

1). to reconstruct the raster image I

2). to manipulate the preimage by geometric transformations, which do not change straight lines, e.g. by affine transformations.

The reconstruction of I is done in two steps. At first every pair of points, representing the start and the end of an ESS will be connected by the corresponding DSS. In this way we achieve an identical reconstruction of the boundaries S_i /5//6/.

The second step consists of the filling of the inside of the regions with the corresponding gray level. The reconstruction algorithm was realized on the A6472 image processing system for images of the size 256x256x8 bits. The filling algorithm works in parallel with a time complexity of O(n/8), where n denotes the number of different gray levels for the region labeling. The arrived images have been proven to be identically to the images which from they were built up. The affine transformation will be realized by a single transformation of the coordinates of the 0-cells without any loss of information. In figure 4 the result of the rotation on the list structure is shown. Comparing the rotated images in Fig. 2-4 we can see, that the number of the digital straight segments of the boundary has increased in Fig. 2 and Fig. 3. There are 24 DSS as in Fig. 2 as in Fig. 3 against 8 DSS in the origin. The locations of the degradation of the straight segment are enclosed in rectangles. In Fig. 4 we have also 8 DSS therefore no any degradation occurs. In this way we are able to adapt the list structure geometrically to other image fragments. Multiple transformations can be done because for the list structure the principle of superposition is satisfied. If we restrict the transformations to the superposition of rotation, translation and reflection we can realize a reconstruction of the transformed raster image without any disturbances of the image structure.

References.

/1/ Kempe V. et al.: Interaktives Bildverarbeitungssystem robotron A6472. "Neue Technik im Büro", 26(1982)5 131-137

/2/ Kovalevsky V.,Luth N.: Parallel algorithms in image processing and computer graphics. "parcella'86",Akademie-Verlag,Berlin,1986

/3/ Kovalevsky V.: Structural image analysis. Eight International Conference on Pattern Recognition, Paris, Oct., 1986".

/4/ Kovalevsky V.: The Topology of Cellular Complexes as Applied to Image Processing. In this issue.

/5/ Kovalevsky V., Wieczorek A.: Fast Recognition of Digital Straight Segments. Technical Report TR 86-73-03 Central Institute of Cybernetics and Information Processes Berlin, GDR, 1986

/6/ Anderson T.A., Kim Ch.E.: Representation of Digital Line Segments and Their Preimages. Computer Vision, Graphics and Image Processing, 30(1985) 279-288.

IMAGE PROCESSING ALGORITHMS ON A TRADITIONAL COMPONENTS BASED CELLPROCESSOR

P. Köles[1)]

1. Introduction

A cellprocessor is a highly parallel computing unit which is especially suitable for the parallel computation of image processing problems. Most of them need thousands of the same local operations which can be solved in parallel by cellprocessors.

In Szeged different cellprocessor architectures have been designed [8-10]. One of them is a so-called cellprocessor ROLL which is based on traditional catalog circuits [9]. Because of measurement problems a small cellular space can be built from traditional components. Therefore this cellprocessor is intended for connecting to a mini or microcomputer increasing its performance by 50-200.

For this architecture a set of cellular image processing algorithms has been designed. Most of algorithms are image preprocessing algorithms but there are a few higher level algorithms, too. Not only image processing, but fixed point arithmetic and Boolean operations have been also designed and tested.

Section 4 gives a short sketch of the most characteristic algorithms of this set of operators.

2. Cellprocessor architecture

One way of implementation of cellprocessors is to make special integrated circuits containing tens or hundreds of cells, and then to build cellular space or cellprocessor from them.

Another way of implementation of cellprocessors is to build cells (and cellprocessor) from standard (catalog) circuits. Such cellprocessors are called traditional component cellprocessors. Such a cellprocessor has been designed. The cellprocessor ROLL as well as the other cellprocessors consists of three main parts [6], namely the cellular space (CS), the cellular space control unit (CCU) and the input-output processor (IOP) (Fig. 1.).

The cellular space consists of 240 2-state cells rearranged into a matrix array of size 15x16. A standard component cell needs a few packages, so in a given volume we can build much smaller cellular space from

1) Research Group on Automata Theory of Hungarian Academy of Sciences H-6720 Szeged, Somogyi u. 7., Hungary

catalog circuits than from custom-designed LSI circuits. But the standard component cells are very fast, because they store their transition functions in a memory of their own. So a global transition step of an algorithm needs a clocktime only.

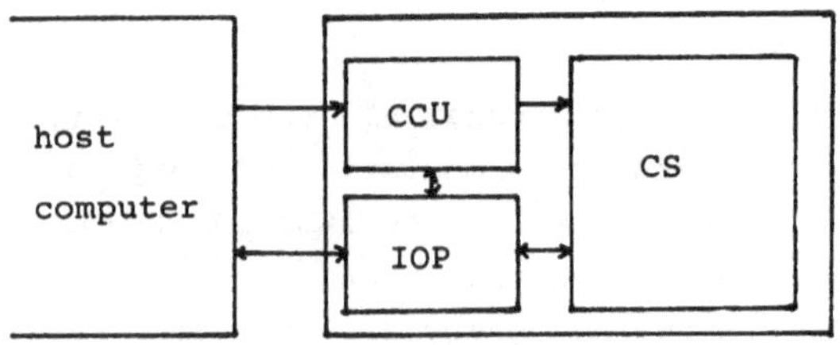

Fig. 1.

Using this small physical cellular space we can emulate a logically larger cellular space without any overhead.

CCU is a traditional microprogrammed control unit which controls the cellular space. IOP moves data from the host computer to the cellular space and vice versa. It also has a buffer memory. IOP and CCU can work in parallel.

The architecture is reviewed in more details in [9].

3. Special abilities of the cellprocessor ROLL

Because of the very small cellular space there are special abilities of architectures aided the effectual programming.

In most cases a small (part of a) processing element (PE) may be tailored in a very sophisticated manner into the cellular space. In the cellular space there is no room to produce input data of PE in due course and format. Therefore there are some special access mode of the element of the IOP buffer memory. For instance for a window operation the necessary pixels may be reordered into the most appropriate sequence for the input of PE. To produce the best geometrical format of input data bits, there is an additional layer over the cellular space. Data bits can be written into the elements of this layer in a whatever manner. The layer shifts in one direction and is connected to the top and the bottom of the cellular space.

Beside the only one external bit, cells have five internal bits, as well. One of them may be handled dinamically, depending on the actual computation. It increases a little bit the computing capacity of the cells. If a cellular algorithm requires, all the five internal bits may be changed under a generalized line. This property makes the emulation of larger cellspaces possible.

4. Cellular operator set

The very small size of the cellular space determines the possible cellular image processing methods. From the three different basic methods (zero-, one-, two-dimensional processing [1]) we can use the zero dimensional (OD) processing only. In this case the size of the cellular space does not depend on the size of the image, operations are performed row by row on the image.

More than two dozens of cellular algorithms have been designed and tested for the architecture. Most of them are basic image preprocessing algorithms such as convolution, different filterings, gray-scale transformation and so on. There exist higher level algorithms, for example histogramming and pattern matching. The available operator set is expandable depending on the request of endusers.

Simple geometrical transformations (reflections, transposition, moving, special rotations) are traced back to changing of different pairs of rows and columns. Specially, rows (columns) may be in diagonal directions, assuming that the host computer or the display processor can handle it. In this case one or two counters are embedded into the cellular space, which give the pointer pairs of rows or columns to be changed. Enlargement can be performed directly in the cellular space.

For convolution and linear gray-scale transformations the main part of PE is a constant multiplier. In case of convolution the coefficients of the mask are encoded into a sequence of internal state bits, so changing the internal states before every pixel determined by the sequence, the different multiplications of a given mask may be performed in one multiplier one after the other. [5] details this algorithm.

Although the Laplacian and Sobel transformations are also convolution type algorithms, because of the special coefficients of their mask, more convenient and faster solutions can be given for them [2]. They use the rows of cellular space for shifting to the left (multiplication by two), for multiplication by -1, for summarizing and so on. Fig. 2. shows the sketch of cellular algorithm for Laplacian transformation.

PE for pattern matching is a bit complicated, but very similar to the PE of Fig. 2. It can perform the operations

$$\max_i |x_i - p_i| \text{ or } \sum_i |x_i - p_i|$$

where x_i is a pixel, p_i is the appropriate element of the given pattern.

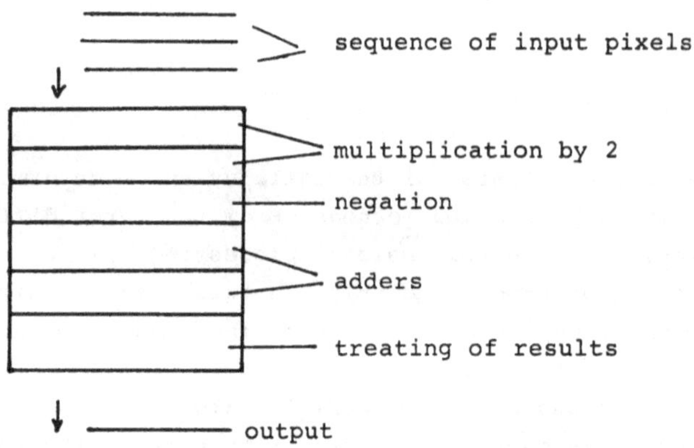

Fig. 2.

PE for median-filtering consists of special sorting rows. Pixels under the given window are sorted while they are shifted through the cellular space, so we can choose the output pixel from a line of the shifting layer in the appropriate time step.

Histogramming must be partitioned, because only seven counters may be tailored into the cellular space (Fig. 3.).

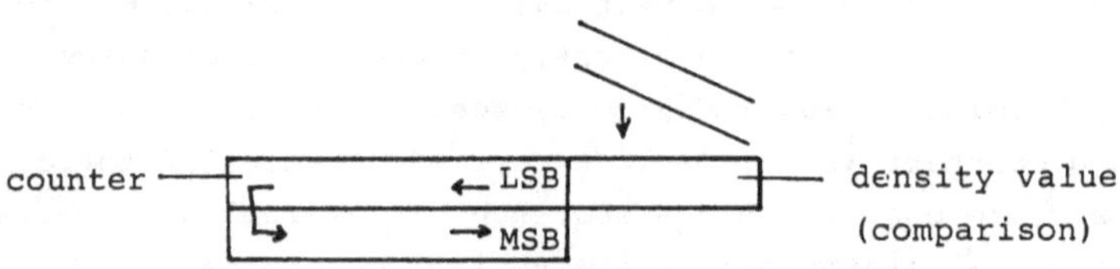

Fig. 3.

Possible pixel densities are encoded into internal bits. The counter is a simple carry-sum adder, where sum bits are stored in the dynamic internal bits. To preclude overflow counters must be curved. When all the pixels have been shifted through the cellular space, the content of counters are also shifted out, internal bits are changed for storing new density values and the procedure is started again. The execution time of this algorithm is not so good than the time of other ones, because of the several phases of procedure. (Each phase needs a total image transmission between the cellprocessor and host computer.)

In spite of the fact that the cellprocessor ROLL has been designed mainly for simple image processing tasks, it is universal, so it is suitable for performing arithmetic and Boolean operations, too. A few algorithms of these topics have been designed and tested, such as sorting, addition and multiplication of fixed point vectors, evaluation of Boolean expressions (for process control), and so cn.

5. Estimated performances

In this point some estimated execution times are shown. Because of the stored transition functions, one global transition step of the total cellular space is a clocktime only (8 MHz). But the parallelism is not so high because of the small cellular space. For the estimation we assume images with about a hundred thousand six-bit pixels. In most cases the real execution time is determined by the time of image transmission to and from the cellprocessor. It is almost one second. Usually it can be overlapped with the execution time of algorithms. Table 1. shows the estimated pure computational time of cellular algorithms.

We emphasize once more that this cellprocessor is intended to be a peripherial device of mini or microcomputers. In this case the system performance may be increased by 50-200.

Algorithm	Time (sec.)
Laplacian transformation	0,225
Convolution (3x3 window)	0,675
Convolution (5x5 window)	1,875
Median-filtering (3x3 window)	0,563
Pattern matching (10x10 pattern)	5
Boolean operations	500 Mops

Table 1.

References

[1] Katona, E.: Computation of window operations by cellular processors. Proc. Methoden und Algorithmen der digitalen Bildverarbeitung, T. U. Dresden, 1985, pp. 17-25.

[2] Kovács, I.: Cellular program for the Sobel transformation on a traditional component based cellprocessor. Proc. PARCELLA 84. Akademie Verlag, Berlin, 1985.

[3] Köles, P.: Picture processing algorithms on large size cellular spaces. Proc. Automatische Bildverarbeitung, Kammer der Technik of GDR, 1985.

[4] Köles, P., Katona, E.: Cellular solution methods for different classes of picture processing algorithms. Proc. Parallel Computing 85, W-Berlin, North-Holland, 1986, pp. 229-234.
[5] Köles, P.: Picture processing on different types of cellular processors. Proc. PARCELLA 86. Akademie Verlag, Berlin, 1986, pp. 114-121.
[6] Legendi, T.: Cellprocessors in computer architecture. Computational Linguistics and Computer Languages 11, 1977, pp. 147-167.
[7] Pratt, W. K.: Digital image processing. J. Wiley and Sons, 1978.
[8] Tóth J.: A 64 Kcell LSI based cellprocessor prototype (SPP). Proc. PARCELLA 84, Akademie Verlag, Berlin, 1985, pp. 154-161.
[9] Zsótér, A.: Cellprocessors based on traditional components. Proc. PARCELLA 84, Akademie Verlag, Berlin, 1985, pp. 162-166.
[10] Zsótér, A., Legendi, T.: A 16-state cellprocessor. Proc. Parallel Computing 85, W-Berlin, North-Holland, 1986, pp. 509-514.

EXPERT SYSTEMS - A MODERN SOFTWARE TECHNOLOGY

REINHARD KLETTE

Abstract. Expert systems have demonstrated that Artificial Intelligence (AI) may be applied for solving concrete problems. So far, expert systems are successfully operating in readily understood (small) areas charaсteized by typical knowledge structure and inference mechanism s, mostly of heuristic type, where well-build expert knowledge is available by cooperative experts, literature etc., where good possibilities for testing are given, and where I/O may be verbalized by simple means. Also in Computer Vision, an increasing amount of research projects is related to expert, or knowledge - based systems. In this paper, after a short introduction to expert systems some concrete projects related to Computer Vision are briefly described. A few fundamental topics in logic are mentioned to demonstrate the principal difference between classical theories and the treatment of "common-sense knowledge" by computers. The emphasis at this point is on the principal importance of concepts of logic for the design of expert systems. Finally, some essential topics of further research are listed.

1. Introduction

New technologies are established for rationalization to save expenses, or for creating extended possibilities for solving problems that could'nt be solved before. Expectations in the new software technology of expert systems did raise up in relation to both of these aspects based on a few examples of expert systems as, e.g. PROSPECTOR, MYCIN or R1 (5) that have gained a very high attention. For example, the "most extensive" expert system so far, R1/XCON (6), containing more than 5000 rules is one of the decoys that lure into this field. Designed for reducing the number of incorrect or misunderstandable orders to Digital Equipment it came out that its benefit is much more than originally expected. Based on the exact and fast work of R1/XCON, now computer systems may be configured at the site of the customer. Recently DEC is claiming that R1/XCON is saving 10 millions of Dollars per year. For example, the construction of a Final Assembly and Test Plant in USA could be cancelled, and, as a side-effect, the new language OPS5 was implemented which is sold now as a tool for creating rule-based expert systems. In relation to such impressive reports, the impact of expert systems upon the progress in Computer Vision is much more modest (18). So far, no expert system in Computer Vision may be called an operational one, although some are close to this state, e.g. (10, 13, 20).

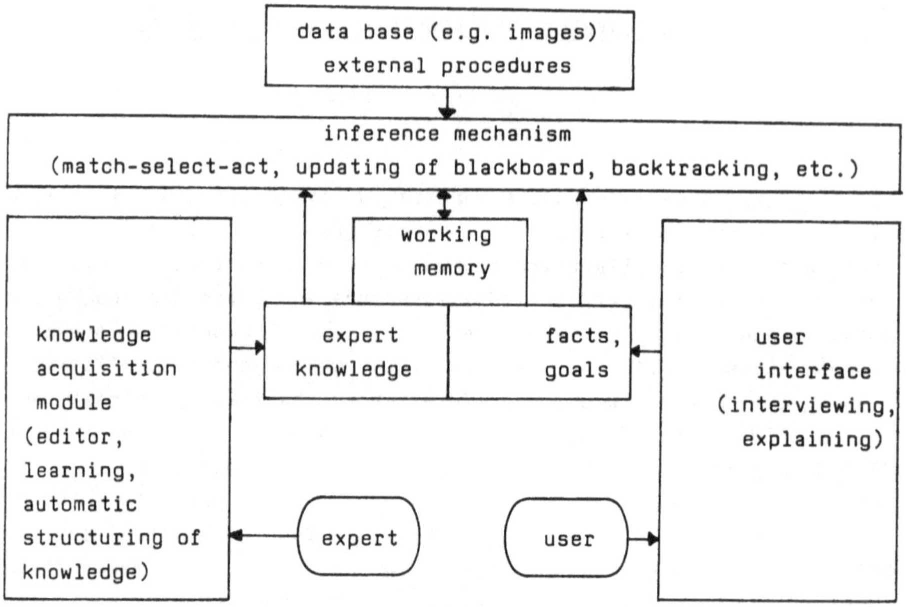

Fig.1. Shell and knowledge base of an expert system: The shell (inference mechanism, knowledge acquisition module, user interface) may be considered as an "empty expert system" which is initialized by filling the long-term memory with expert knowledge. Facts or goals (short-term memory) and intermediate results or alternatives (working memory) are generated and stored in the knowledge base during one session with the expert system. Arrows may be read as "has influence to".

An **expert system** (5, 21) is a computer program that emulates expert knowledge by applying its inference methods to a knowledge base. Thus, expert systems are characterized by a qualitative goal (to cope with the expert perfomance) as well as a basic architecture (inference module and knowledge base). Because in Computer Vision the "expert knowledge" sometimes is of very lew degree (e.g., interpretations of 3D scenes may be done by infants already) also the notation knowledge - based system is in wide use.

The separation of programs and data was the starting point of the development of data managing techniques as data structures or relational database systems. By a further separation of programs into control structure and specific knowledge of the area of application, an even higher modularity of software products was made possible. (The knowledge about the appropriate control or inference mechanism may be considered to be meta-knowledge in relation to the specific knowledge of the area of application.) On this way of rationalization of software production, not only different types of control or inference mechanism s and knowledge representations could be studied and placed at disposal, e.g. (5,

9, 15, 16, 21), but also subsequent modules for knowledge acquisition, or user interfases for interviewing or explaining (Fig. 1). It is mainly this engineering point of view which is speeding - up the development in expert systems, and less the academic interest in turning semantic theories into working programs (for solving problems which are considered to be not solvable so far).

To outline a second analogy, as well as database technology has led to new applications of computers also expert system technology will broaden the spectrum of tasks solvable by computers in practice. But, nevertheless these new technologies don't replace conventional programing, they have to be based on a high standard of computing characterized by large memory capacity and fast processing (especially for designing expert systems by "fast prototyping") and developed software environments, and they are integrating traditional programing. To make a rough guess, operational expert systems will consist of AI software to about 20 % (what might be called the"essential core" of the system) only, because modules for graphics, image processing, process control, statistics, arithmetic procedures, user dialogue etc. will be written in traditional languages. Expert systems cannot be used for equalizing being behind in traditional computing.

2. Design and Implementation

The specific properties of the field of application are important for specifying the features of the expert system. This sentence is not as trivial as it looks at the first glance because some years ago the idea of a "general problem solver" did have some followers. For designing an expert system at first the problems which have to be solved and the solution methods which will be used have to be brought into correspondence. During the short history of expert systems it came out that such modular structured computer programs may be succesfully used in areas characterized by "diffuse" knowledge where the application of computers was considered not to be appropriate so far. Now, expert systems are preferably designed for such areas of "bad structured knowledge" where no general algorithm is known for a closed treatment of the given class of problems. For example, for the task of selecting a specific method for edge detection for a given class of pictures, no method stands out as the best in general, and there has been no convergence among the many techniques proposed (although the Marr - Hildreth approach based on the zero crossing of the second derivative of the intensity gradient appears promising). Thus, an expert system approach may be used for selecting a "best" edge detection method by following "rules of thumb" about the qualitative evaluation of the different combinations of methods, and results obtained by these methods, or about possible combinations of edge detection methods and subsequent "improvements" of the given edge segments. Such a system would be a typical consultant - it doesn't solve the given image processing problem in the sense of a procedure that may be applied for edge detection, but it outputs one, or some weighted proposals for such a procedure. Examples of expert systems in this direction are described in (11, 14) where the more general problem of image segmentation is dealt with. The approach in (11) is a typical rule-oriented one (Fig. 2), whereas an object - oriented approach (15) was prefered in (14), cp. Fig. 3.

In general, expert systems may be classified according to problem classes. Systems for diagnosis or classification are studied most profoundly. Here, starting with given symptoms or measured values (in the short-term memory in Fig. 1) a certain recognition of an object, pattern, situation etc. has to be realized where the object, pattern, situation etc. was known to the system before by its content of the knowledge base. For example, in ACRONYM (3) models of 3D objects are used for comparisons. Systems for construction of solutions may be characterized to be either design or planing systems. In the first case, objects have to be constructed which statisfy certain constraints. In the second case, a sequence of actions has to be constructed which allows to obtain a goal state, i.e., in this case start and goal state are given initially in the short - term memory, and intermediate states are generated in the working memory. For example, in XAMBA (10) the start state is characterized by an informal description of the task which a picture analysis program has to solve, and the goal state is characterized by the request that this program has to be coded in the language LAMBA. Intermediate states represent this program at different levels of refinement. Finally, systems for simulation are transforming an initial configuration into possible subsequent configurations, and may be subprocesses of systems for construction for testing possible consequences or alternatives.

After selecting the type of knowledge representation (which is formally equivalent to the specification of the system's language), and the inference mechanism for reasoning (which is formally equivalent to the specification of a derivability relation for infering knowledge in the system's language), the normally expected comfort of an expert system may be added by improving the shell of the system. In most cases, the knowledge acquisition module may be characterized to be the "bottleneck" of the system, especially in the period of cooperation between the knowledge engineer (who designs and implements the expert system) and the expert of the field of application (who places at disposal his expert knowledge). The basic problem of the knowledge acquisition module is the need for a well-developed knowledge base, offering lots of different types of relations between units of knowledge, for emulating complex knowledge but the difficulty of having the actual content of the knowledge base in mind during altering it. Automatic structuring of knowledge, consistency testing, learning are three key issues to this situation. If the knowledge engineer starts himself to input expert knowledge into the system this normally signals the dead of the project.

Similarily, the user interface decides essentially whether the user will accept the system or not. In general, an expert system should have at least some of the following features simultaneously:

(i) it understands problems formulated in its area,
(ii) it may rate its own competence to such problems,
(iii)it may give hints to related areas,
(iv) it explains its state (i.e. intermediate solutions, alternatives, reasons for making desicions etc.) upon request,
(v) the content of the knowledge base may be altered for changing the systems behavior,and
(vi) there are some service modules for handling the knowledge base.

Recent expert systems are still far behind an optimal appearence of all of these features.For example, the formulation of problems by the user is restricted to."typical vocabularies" (a Natural Language Interface is yet a research issue), and automatic learning of knowledge is still in its beginning in relation to its practical relevance.
For implementing expert systems an increasing manifold of tools is under development. Normally, prototype systems are written in typical AI languages as LISP dialects (22) or PROLOG (4), and in case of operational systems there is still some preference of traditional imperative languages. Programing enviroments for implementing expert systems may be characterized at different levels,

(i) available AI languages,
(ii) language extensions for knowledge representation and inference mechanisms,
(iii) shells ("empty expert systems"), or
(iv) complex program packages (as LOOPS, KEE, SRL) for AI programming.

Based on developments of AI hardware, as LISP- or PROLOG-machines, or of some components for attractive design of the user surface, these software tools will speed-up the development and applications of expert systems.
A new technology is propagating in successive periods. First, in a few places it is introduced and succesfully demonstrated. Second, its effects are studied in relation to rationalization or for solving problems which were considered to be unsolvable so far in selected fields of application, where these problems are assumed to be in optimal correspondence with this technology. And third, in the positive case it is integrated into these fields of application.In Computer vision ,this third period already has started for expert systems (1,3,5,18) .

```
Rule (802):
     IF:    (1) the REGION SIZE is LOW
            (2) the ADJACENCY with ANOTHER REGION is HIGH
            (3) the DIFFERENCE in BOTH REGIONS FEATURES is NOT HIGH
     THEN:  (1) MERGE the TWO REGIONS
Rule (1507):
     IF:    (1) the LINE END POINT is OPEN
            (2) the LINE AVERAGE GRADIENT is HIGH
     THEN:  (1) EXTEND the LINE FORWARD
Meta Rule (9):
     IF:    (1) the PREVIOUS PROCESS was FOCUS OF ATTENTION
            (2) the PREVIOUS PROCESS was ACTIVE (a rule did fire)
     THEN:  (1) MATCH the REGION ANALYSIS RULES
```

Fig.2. Production rules for segmentation (11) :The conditions of rule 802 are matched to the properties of the actual region and one of its adjacent regions. If all the three conditions match,the merging of the two regions will be performed.Similarily,rule 1507 is applied for extending lines,and meta-rule 9 for selection of the next process.

3. Expert Systems in Computer Vision

Writing programs in Computer Vision always means to use specific knowledge about the objects which have to be interpreted in the given pictures; Computer Vision without using knowledge would be a contradiction in itsself. But, AI approaches are characterized by making explicit use of knowledge represented in some rule-oriented (Fig.2) or object-oriented (Fig.3 and Fig.4) language, and by a seperated inference strategy operating on such explicitly given units of knowledge.
There are different classes of problems in Computer Vision where expert systems seem to be the appropriate technology. In the sense of the task successively solved by R1/XCON the configuration of Computer Vision systems may be considered. More specific, for a well-developed image processing system the task may be that programs for Computer Vision have to be synthesized based on some informal descriptions of processing goals, or, to be more pretentious, also on some automatic analysis processes of the given pictorial data. Projects in this direction are reported in (10) and (20), where the commercial image processing system AMBA and the transportable software package SPIDER were used as basis, respectively. In these approaches, the emulated expert knows the specific possibilities of the given image processing system as the effects of procedures in dependece of parameter specification, the ways to connect such procedures for solving subtasks of image analysis, some "tricky" solutions to standard subtasks etc. In the approach of (10), a hierarchical refinement process (Fig. 5) was implemented where a hierarchy of program graphs Pi is stored in the working memory (Fig. 1) for repre-

senting the actual state, and a specific refinement step is specified by the inference mechanism in relation to user requests stored in the short-term memory.

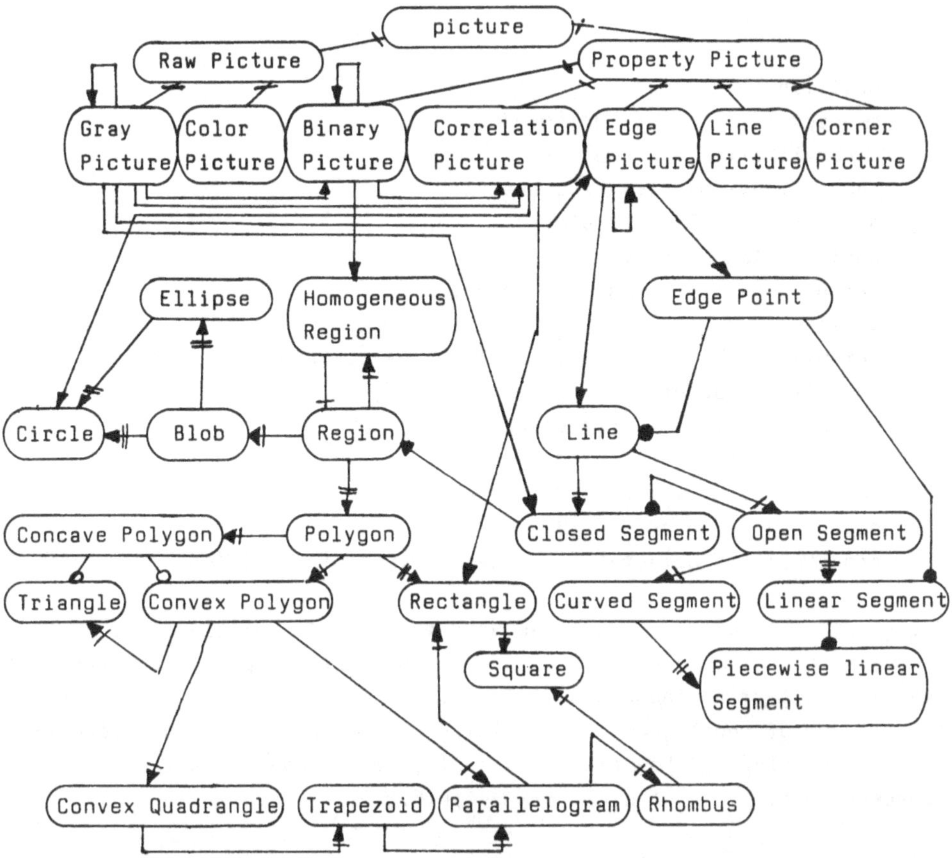

Fig.3. Network knowledge organization (14) for automatic image segmentation: "Image Features" are objects of the network (labels of the vertices) and the different edges represent different types of relations between these objects. For example, ——▶ denotes a "Transfer Process" which is labeled by the concrete process as edge detection, thresholding, linking etc..

Many expert system projects in Computer Vision are directly related to the process of interpreting or analyzing specific pictures by emulating experts of methods or algorithms (e.g. for performing image segmentation on the given class of pictures (11, 14), or for detecting objects of special shapes (7), or experts of the specific field of application (e.g., for a complete medical interpretation of pictures in nuclear medicine (19), or for model-driven interpretations of pictures in remote sensing (13)). In just a few cases, the declared goal is the design of complex Computer Vision systems for understandig a wide range of pictures, as for ACRONYM (3), the VISIONS system (17), or SIGMA (13, Section 3.2).

```
name of concept: string
   number of substructures: integer
   description: text
   position in the network: integers
   concept of: list of instances
   generalization: concept
   specialization: list of concepts
   similarities: list of concepts
   context flag: YES/NO
   semantic part of: list of concepts
   semantic parts:        ┐
   semantic attributes:   ├ complex substructures
   semantic structures:   ┘
   necessary part of: list of concepts
   necessary parts:       ┐
   necessary attributes:  ├ complex substructures
   necessary structures:  ┘
   procedure for computation of certainty factor: string
```

Fig.4. Data structure CONCEPT as used in (19): At this level, a CONCEPT consists of 18 pointers to slots which may be complex substructures itself. The list of instances (fifth slot) is produced bottom-up during the process of model generation. Relations as "generalization","specialization","similarity","semantic part of",and "necessary part of" may be represented by edges of a network as in Fig.3.

In ACRONYM very general methods of representing geometric relations by geometric transforms are used. 3D objects are modeled as hierarchies of generalized cylinders, and geometric relations between objects are described by 3D geometric transformations between object-centered coordinate systems . The inference mechanism of the system is characterized by symbolic reasoning for analyzing scenes including prediction, description, and interpretation of objects. The basic units for object description are frames (15) which are the nodes of a structure called the Object graph. Using knowledge of this object model as well as some feedback to low level processes the system constructs a Prediction graph in the working memory (Fig. 1). Predictable entities are called observables, and in ACRONYM two kinds of observable features could be handled: Ribbons as 2D specializations of generalized cylinders, and ellipses as generators of generalized cylinders. Finally, the Interpreter Module of the system matches these observables and their geometric relations to the Object Graph where measurements of image features are used to con-strain model parameters. In this way, iterations of prediction, description and interpretation are used to identify a maximal subgraph of the Prediction graph that matches the actual situation in the given scene. The system was demonstrated for aerial images of scenes containing three types of passenger jets. But, its real importance is in demonstrating interactions between models, constraints, feedback to low-level processing and interpretation. The modular structure of expert systems was of essential importance for realizing such a complex research project. Nevertheless, it may be argued that derivations of operational systems out of such projects will be more close to traditional programs because modularity is not of that importance for the final system.

parent graph Pi

global data: d2,d3

child graph Pk

local data: d6,d7

term(Pk,e1,Pi,s2)= a3
term(Pk,e2,Pi,s2)= a4

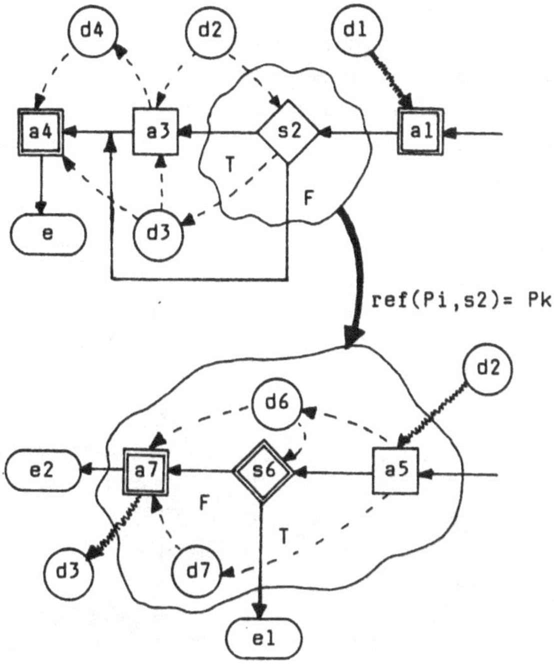

Fig.5. Refinement process for constructing programs at different levels of abstraction (10): Vertices are of type data d1, d2, ..., action a1, a2, ..., split s1, s2, ... or exit e, e1, e2. Any non-terminal action or split vertex (in graphical representation: not double-lined) has to be refined where the inference mechanism is building up functions ref and term. Syntactic restrictions are given by the number and type of global data vertices, and the number of exits. Semantic restrictions are generated top-down.

The VISIONS system is perhaps the most comprehensive expert system today. Basically, this system builds intermediate symbolic representations of the image data using knowledge-free segmentation processes, i.e. without using specific knowledge about the objects in the image. From that intermediate level, partial image interpretations are constructed by associating object labels with intermediate primitives. These labels are used to activate those parts of the knowledge network which are related to the hypothesized objects. Here, rule-based object hypothesis direct further processes at the intermediate level for object matching and object inference (grouping, splitting and adding regions, lines and surfaces).

The SIGMA system (Fig. 6) is based on the idea that an object hypotheses may be strengthened by object hypothesis about image areas which satisfy geometrical relations with the original hypotheses (evidence accumulation). In (13) several open problems are mentioned which should be solved for improving this approach, e.g., managing of failure and exception, meta-knowledge about dependencies of geometric relations, or managing of multiple mutually conflicting interpretations. These problems are, in dead, directed to fundamental research issues of expert systems.

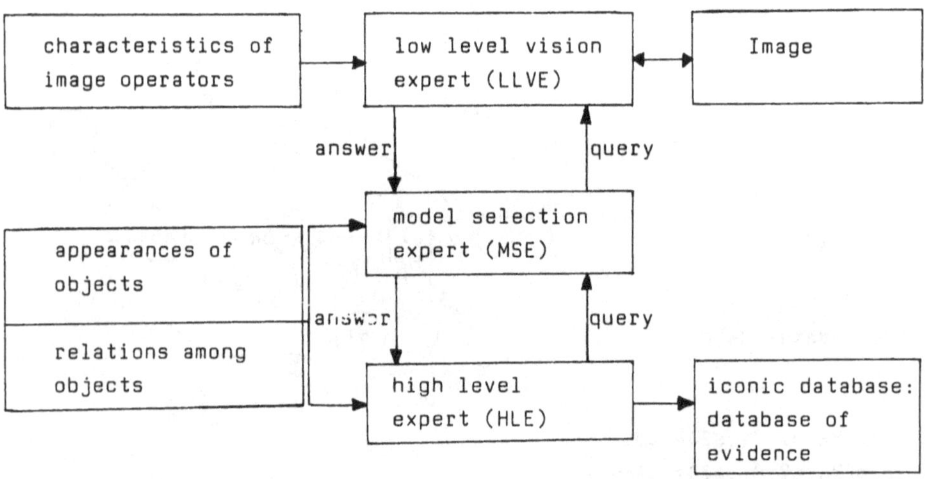

Fig.6. Block diagram of the system SIGMA (13): The high level expert is performing spatial reasoning using evidence accumulation. Queries are given to the model selection and low level vision experts for establishing "situations" which consist of consistent pieces of evidence.

4. Logic Fundamentals

In an abstract sense, any expert, or knowledge based system is defined by a finite set of formulas which may be called the set of axioms (the long - term memory in Fig. 1) and a derivability relation (the inference mechanism in Fig. 1). The system's language may be rule-oriented (Fig. 2) or object-oriented (Fig. 4), where both types of languages may be characterized to be different points of view to first-order predicate calculus (8), and subsets of such languages may be represented as associative networks, cp.(16). A **formula** is an expression in the system's language which is well-build according to the syntactic restrictions of these language. An **inference system** IS is defined by a language L, a knowledge base kb0 which is a finite set of formulas of language L, and a derivability relation $\vdash$ which maps formulas into formulas, e.g. concepts in the sense of Fig. 4 into instances of these concepts. Formally,

$$IS = [L, kb0, \vdash] ,$$

and in the sense of classical logic (8) this may be also called a **theory**. The set kb $= \{w: kb0 \models w\}$ consists of all formulas which may be derived from kb0 by $\vdash$, and $\models$ is the reflexive and transitive closure of relation $\vdash$. So far, an inference system is without any semantic interpretation. At that point, strings as REGION SIZE, LOW, ADJACENCY etc. (cp. Fig. 2) don't have any concrete meaning.

A **domain** is defined by a set of objects or individuals in the specific field of discourse and sets of functions and relations on these objects or individuals. For example, a concrete set of system modules of an image processing system and compatibility relations between these modules would be a domain for the task of configuration of Computer Vision systems. A set of objects of a concrete image analysis situation (region, line, area, integers etc.) and of relations between these objects (adjacent, low, active etc.) is an other example of a domain which corresponds to Fig. 2.

An **interpretation** φ is a unique mapping that maps all constant symbols in formulas of language L onto objects of a given domain, and all predicate symbols of L onto relations of the domain. Altogether,

$$IS_{\varphi} = [L, kb0, \vdash, \varphi]$$

may be called an **interpreted inference system** or a **semantic theory**. For example, by φ(REGION) = region the word REGION will be interpreted by something well-defined by the notion "region", e.g. a finite and connected set of pixels in the sense of 6 - neighborhood having const0 elements at least.

In classical logic (8), a **model** is a semantic theory IS_{φ} where all formulas in the set kb come out to be true, i.e. for $w \in$ kb it is assumed (!) that the truth value of $\varphi(w)$ in the given domain is known. But, for statements during image analysis, for example, it is not possible to decide its truth value in general. Here, **semantic theories of knowledge** (2) may be appropriate where monodic operators K (read: "it is known to be true that"), H (read: "it is hypothesized that"), K1, K2, ... , Kn (read Ki as "it is know to agent i that ... is true") etc. are introduced as well as a set ax of basic axioms for these operators. For example, for all $w \in$ kb0 it should be assumed that K $\varphi(w)$. Then, if

$K(\varphi(w1) \wedge \varphi(w2)) \rightarrow K(\varphi(w1))$ is an axiom of the given semantic theory of knowledge then the specific rule $w1 \wedge w2 \vdash w1$ may be used for infering formulas which are known to be true. Altogether, a sematic theory of knowledge $[L, kb0, \vdash, \varphi, ax]$ may be considered to be the formal analog of an expert system.

Typically, the inference mechanisms of expert systems may be characterized to be an implementation of one of the following basic types of inferences. Deductive inferences may be described by the follwing way of reasoning:

if $\forall x[R(x) \rightarrow P(x)]$ and R(a), then P(a) .

Here, inference of knowledge from knowledge (absolute justification) is possible. In a weaker form,

if (most x)$[R(x) \rightarrow P(x)]$ and R(a), then P(a) ,

this type of justification may be used for strong recommendation. By abductive inferences,

if $\forall x[R(x) \rightarrow P(x)]$ and P(a), then R(a),

hypotheses may be infered as possible causes in diagnostic expert systems, and the hypothesis R(a) may be strengthed by additional abductive inferences. Finally, also inductive inferences,

if R(a) and P(a) , then a certain general law for R(x) and P(x) is infered,

may work if "possible classes of laws" are known in advance, and series of objects, a1, a2, ... are considered for accumulating evidence (12). The concrete assumptions about inference of knowledge may be formalized by using the monadic operators K, H, K1, K2,... defined above.

By performing inferences the epistemic status of formulas, i.e. the presumed truth of a formula in relation to a specified semantic theory of knowledge, may be changed. At the lowest level, belief is any formula of language L where certain theories of belief (formalizing consistence of belief or other assumptions) may be used to characterize restrictions. A hypothesis is a justified belief where justification is relativ to the knowledge base generated so far, and this process is formally represented by the derivation of expressions as $H\varphi(w)$ or $Hi\ \varphi(w)$. Finally, knowledge is justified belief which is known to be true, and formally derived by infering expressions as $K\varphi(w)$ or $Ki\ \varphi(w)$. Note, that we have now a definition of this basic notion of KNOWLEDGE, and in (2) some papers are related to this scetched way of "non - classical extensions" of logic.

At that point, it is possible to define features of knowledge which are essential for specifying the structure of concrete expert systems. Such features, as incompleteness, inconsistency, non-monotonity, inaccurracy, relativity, uncertainty, imprecision, informality, significance, relevance etc. are considered in (2), e.g., and these features are characterizing at some extent the new quality by going from data to knowledge processing. Progress in studying these "non-classical extensions" of logic, and such features of knowledge as listed will have an essential impact on the futher development of expert system technology.

5. Conclusions

In rating the present situation in the field of expert systems, especially in Computer Vision, some critical remarks and important topics for futher research may be listed. Despite of the high attention that expert systems, and AI in general have gained during the last years it may be said that this field is still at its beginning. As the difficulties connected with Computer Vision were highly underestimated some years ago at many places (and now, it seems, we do have a time of decreasing interest in this field what is the normal result of this underestimation), now the same is true for expert systems. There should be much more concrete projects in this field for stimulating real progress, and less papers "just talking about it".

Research in the field of expert systems will make contributions to our knowledge about KNOWLEDGE, i.e. about the structur e of inference systems which may be used for certain types of reasoning as well as about features of knowledge as mentioned at the end of Section 4, and how to deal with knowledge which is characterized by these features. So far, expert systems in Computer Vision have demonstrated that different types of inference systems (several agents) have to cooperate for solving complex problems (as illustrated in Fig.6 for three agents) where the iteration between low-level and high -level processes seems to be essential. Only close to the configuration or program synthesizing area there is some hope that operational systems may be possible in the near future. General research issues of expert systems are, for example:

(i) Learning for simplifying the knowledge acquisition processes,

(ii) explainations of "why" this solution path was selected, and not "in what sequence of steps" the solution was generated so far,

(iii) generalisations of specifi ed expert system approaches,

(iv) inclusion of causal knowledge of the given domain into the inference system,

(v) more comfort at the user surface (Natural language Interface, specialized Hardware/Software components etc.), and

(vi) judgement of its own competence, or the development of own strategies which may be (partially) in conflict to the user's requests.

References

1. D.H.Ballard, C.M.Brown, Computer Vision. Englewood Cliffs,N.J., Prentice-Hall, 1982.
2. W.Bibel, Ph.Jorrand (eds.), Fundamentals of Artificial Intelligence. LNCS 232, Springer, Berlin Heidelberg New York London Paris Tokyo, 1985.
3. T.O.Binford, Survey of model-based image analysis systems. Int. J. of Robotics Research 1(1982), 18-64.
4. W.Clocksin, C.Mellish, Programming in Prolog. Springer, Berlin Heidelberg New York , 1981.
5. P.R.Cohen, E.A.Feigenbaum, The Handbook of Artificial Intelligence. Heuristech Press, Stanford, C.A., 1982.
6. J.Mc Dermott, R1 revisited:four years in the trenches. AI Magazin 5(1984)3, 21-32.
7. M.Ender, C.-E.Liedtke, Repräsentation der relevanten Wissensinhalte in einem selbstadaptierten Bilddeutungssystem. IFB 125 (8.DAGM - Symp.), Springer-Verlag 1986, 219-223.
8. A.Grzegorczyk, An Outline of Mathematical Logic. D.Reidel Publ. Company Dordrecht - Holland/Boston - USA, 1974.
9. B.Hayes-Roth, A blackboard architecture for control. Artificial Intelligence 26 (1985), No. 3.
10. R.Hesse, R.Klein, R.Klette, XAMBA - A knowledge-based programming environment for automated picture analysis. Proc.Int.Conf. "Systems of knowledge and picture processing", Smolenice, Czechoslovakia, 24 -28 Nov.1986, 105-109.
11. M.D.Levin, A.B.Nazif, Rule-based image segmentation. Computer Vision, Graphics, and Image Processing 32 (1985), 104-126.
12. R.Klette, R.Wiehagen, Research in the theory of inductive inference by GDR mathematicians - a survey. Information Sciences, 22(1980), 149-169.
13. T.Matsuyama, Knowledge organization and control structure for image understanding. Proc.7th Int.Conf.on Pattern Recognition, July 30 - August 2, 1984, Montreal, IEEE, 1118-1127.
14. T.Matsuyama, M.Ozuki, LLVE: An Expert system for top-down image segmentation. Journal of IPS Japan 27(1986), 191-204 (in Japanese).
15. M.Minsky, A framework for representing knowledge. in: Psychology of Computer Vision (P.H. Winston, ed.), Mc-Graw Hill, New York, 1975.
16. N.Nilsson, Principles of Artificial Intelligence. Springer, Berlin Heidelberg New York, 1980.
17. E.M.Riseman, A.R. Hanson, A methodology for the development of general knowledge-based vision systems. in: Wissensbasierte Systeme (W. Brauer, B.Radig, Hrsgb.), Informatik-Fachbericht 112, Springer-Verlag Berlin Heidelberg New York Tokyo, 1985, 257 - 288.
18. A. Rosenfeld, Image analysis: problems, progress and prospects. Pattern Recognition 17 (1984), 3-12.
19. G.Sagerer, Darstellung und Nutzung von Expertenwissen für ein Bildanalysesystem. IFB 104, Springer-Verlag 1985.
20. K.Sakane, H.Tamura, Automatic generation of image processing programs.Proc. of CVPR, 1985, 189-192.
21. M.Stefik, J.Aikins, R.Balzer, J.Benoit, L.Birnbaum, F.Hayes-Roth, E.Sacerdoti, The organization of expert systems, a Tutorial. Artificial Intelligence 18(1982), 135.
22. P.H.Winston, B.K.P.Horn, LISP. Addison-Wesley Publish.Comp., Reading, Mass., 1981.

UTILIZATION OF KNOWLEDGE CONTENTS FOR THE AUTOMATED ADAPTATION OF AN INDUSTRIAL VISION SYSTEM

C.-E. Liedtke, M. Ender

Institut für Theoretische Nachrichtentechnik und Informationsverarbeitung, Universität Hannover
Callinstr. 32, D 3000 Hannover 1, FRG

1. INTRODUCTION

A major problem in the use of vision systems in industrial automation becomes the adaptation of these systems to new objects and object properties, changes in illumination, and changes of the camera position. Up to now the adaptation requires the help of experts in the area of image processing and pattern recognition. Since these experts are highly expensive and since their number is quite limited the question arises if it is possible to overcome this situation by entering a part of this expert knowledge into the vision system itself, so that the system can with little support by the user adapt itself automatically to new scene contents.

Since we cannot solve this problem in all generality, we limited ourselves to a very specific area of application, namely the recognition of 2D- and 3D-objects and their position from a single standard TV camera image. In order to simplify this problem further, we restricted ourselves to a class of industrial scenes, where the relevant information about the object type, position, and orientation is contained in the spatial relation of line structures like edges, vertices, circles, and ellipses. We will call these line structures in the following "scene domain cues".

A concept for a fast and flexible image analysis system has been developed under the following considerations: (a) Many similar images have to be analyzed and the analysis procedure must be extremely fast. (b) The adaptation is needed less frequently, and therefore considerably more time is available. The concept is illustrated in Fig.1. During a processing phase a fast algorithm is used for the image interpretation. This algorithm is automatically generated in a preceding adaptation phase by some inference mechanism, which employs on one hand knowledge contents about the object domain of immediate interest and on the other hand knowledge contents about image analysis algorithms.

2. CONCEPT OF THE AUTOMATED ADAPTATION SYSTEM

The concept of the adaptation procedure is shown in Fig.2. The knowledge contents about the object domain of immediate interest are offered to the system in the form of a reference image and a corresponding reference description. From the reference description the system can derive automatically ideal intermediate processing results. The actual intermediate processing results are obtained by processing the reference image. The differences between the actual processed results and the ideal results are evaluated resulting in modifications of the processing path. This procedure is periodically repeated in a processing-evaluation-modification cycle until a pregiven termination criterion has been fulfilled.

Parts of the system have been described in the literature [1 - 3]. In the following the aspect of knowledge representation and coding, which is essential for the solution of the problem is discussed in detail.

3. CODING METHODS FOR KNOWLEDGE CONTENTS

Any knowledge content can be expressed by a declaration or a procedure. For example the knowledge content about the shape of a particular object can be expressed by the declaration:

"The object is round" (3.1)

The same knowledge content could be expressed by the procedure:

"Calculate the center of gravity of the object. Measure the distance between any point of the object and the center of gravity. If all distances remain within a pregiven tolerance interval, then the object is round." (3.2)

Though the knowledge contents in both examples are the same, significant differences become apparent in the use of the representational schemes. One difference refers to flexibility in respect to modification of the knowledge content. If an object in an image appears to be angular, the attribute "round" in the declaration 3.1 needs only to be replaced by "angular" to meet the new situation. In contrast it is quite difficult to adapt the procedure 3.2 to the description of an angular shape. The procedure 3.2 has completely to be rewritten. From this can be concluded, that declarative knowledge representations will find use in applications, where the knowledge contents have frequently to be modified.

A second difference concerns the efficiency in the use of the knowledge contents. The procedural knowledge representation has the advantage, that it contains a description of how to check the validity of the knowledge content. The description has been mostly formulated such, that the validity of the knowledge content can be checked very efficiently. In contrast, the declarative knowledge representation gives no indication, how to check the validity of the knowledge content. Therefore more general and usually inefficient methods are used. In applications, where large amounts of data have to be processed and where flexibility is of minor importance, procedural representations for knowledge contents will therefore be prefered.

Knowledge contents are abstract. The knowledge contents, which a human uses for reasoning purposes are usually represented in a different form than those in technical systems due to the different properties of the systems and the environments in which they work. This situation is described by the term "implicite knowledge representation". In case the knowledge contents are represented in technical systems in a similar form as humans use them, the knowledge representation is called to be "explicite". The use of explicite knowledge representation in machines has the advantage, that the human operator can more easily understand and control technical systems.

The combinations of declarative vs. procedural and implicite vs. explicite representation have led to mainly four coding methods for knowledge contents, namely ALGORITHM, PARAMETER, RULE, and SEMANTIC NET. The different properties of the coding methods are indicated in Fig.3. ALGORITHM and PARAMETER represent the classical and up to now most efficient tools for the formulation and manipulation of knowledge contents. They are strongly supported by algorithmic programming languages like FORTRAN, PASCAL, C, LISP, etc.. There wide use is caused mainly by the technical properties of today's existing computer systems.

The coding methods RULE and SEMANTIC NET permit the explicite representation of knowledge contents. For their use they require an interpretation mechanism. Explicite knowledge representation is mostly supported by expert systems.

The methods of image analysis can usually be subdivided into three groups of processing methods: (1) Low-level-processing, (2) extraction of a first symbolic description, and (3) high-level-processing. Low-level processing includes preprocessing and image segmentation. The extraction of a first symbolic description refers in our application to the extraction of scene domain cues. The processing methods (1) and (2) have in common, that they work on the data structure of pixel arrays, which contain usually a large amount of data. In this group mainly high efficiency and only to a lower degree flexibility is required to handle the large data volumes. Therefore implicite and procedural knowledge representation schemes are prefered. In the group (3) of high-level-processing, where hypotheses are set up, plans are generated and inference mechanisms are applied, the request for flexibility and explicite knowledge representation becomes dominant. The processing methods work mostly on reduced data sets of symbolic data, which employ more general and flexible data structures like trees or graphs.

4. CODING OF KNOWLEDGE CONTENTS IN THE AUTOMATED ADAPTATION SYSTEM

The solution of the adaptation problem, which has been stated above, requires the utilization of knowledge contents (KC), in the following categories:

(a) KC about the object domain

Prior to the adaptation phase the operator has to provide the system with a relevant description of the workpiece, which has to be recognized during the production phase, and a reference description of the reference scene, to which the system will be adapted.

(b) KC concerning processing methods

Image interpretation is a multistep process. The relevant information is enhanced step by step by applying a sequence of elementary processing methods. The KC about the enhancement of the relevant information are contained in the set of elementary processing methods. In connection with the given application this includes KC about the preprocessing, segmentation, the extraction of scene domain cues as first symbolic description, and about grouping methods for object recognition.

(c) KC about evaluation methods

The evaluation is based on the calculation of quality measures and the interpretation of the values of these measures. Since the evaluation is applied sequentially to different levels of abstraction, a global optimization requires the use of consistent quality measures. Consistency is obtained by stating the common properties of the quality measures on a higher level.

Essential KC are contained in

- the properties of the quality measures
- the interpretation of the values of the quality measures

(d) <u>KC about the adaptation strategy</u>

The adaptation strategy employs heuristic KC about the methodology of adaptation, the use of the evaluation methods and KC about the context for which adaptation rules are valid.

(e) <u>KC about the system control</u>

In order to process the data KC are needed, which describe how to run the actual processing system. These include KC about

- the functioning of the expert system shell
- the communication paths of the subsystems
- the mechanisms of rule selection
- the elementary processing procedures
- the evaluation procedures
- the individual rules
- the activation of the processing and the evaluation system

(f) <u>General KC</u>

All KC, which have been mentioned so far are specific for the object domain, the area of application, or the processing system. In addition some general KC are needed, like KC about the laws and rules of mathematics and logic, KC about the meaning of general terms, like "edges", "vertices", "circles", "ellipses", and KC about general procedures like minimation, maximation, etc..

(g) <u>KC about the processing algorithm</u>

The KC about

- the optimal sequence of elementary procedures and
- the optimal parameter values

represent the results of the adaptation process. These KC have been obtained by the inference mechanism of the system from the other KC which have been mentioned above.

The coding methods for the KC are listed in the table of Fig.4. KC which are either related to the processing of large volumes of data or which are used frequently are coded for reasons of efficiency by the scheme ALGORITHM. Heuristic KC about the interpretation of the values of the quality measures, about methods and methodologies require an explicite - procedural representation and are coded by the scheme RULE. KC, which have frequently to be exchanged, like the KC about the object domain, which changes with each application or KC about the processing methods, the evaluation methods and the rules, which are frequently updated and modified during the adaptation process need an explicite - declarative representation. They have been coded by the scheme SEMANTIC NET.

Examples for the different representation and coding methods ALGORTITHM, RULE, and SEMANTIC NET as they have been used in the proposed system are shown in Fig.5 - 7.

5. RESULTS

The described knowledge based vision system has been simulated on a VAX 11/780. The symbolic data base contains presently about 40 nodes of a semantic net and about 60 production rules. An example for the stepwise adaptation in connection with the recognition of vertices of an angle iron is shown in Fig.8. The adaptation process needed about 20 minutes of CPU-time, where about 10% were used by the expert system, about 90% by the image processing system and less than 1% by the internal communication between the expert system and the image processing part.

6. REFERENCES

1. Liedtke, C.-E.; Ender, M.; Heuser, M.: Komponenten eines adaptiven wissensbasierten Bildverarbeitungssystems zur Lageerkennung von Objekten, 7. DAGM Symposium "Mustererkennung 1985", Erlangen, 1985.

2. Ender, M.; Liedtke, C.-E.: Evaluation Criteria for Assessment of Processing Quality in a Fully Automatic Knowledge Based Vision System, EUSIPCO-86, The Hague, 1986.

3. Liedtke, C.-E.; Ender, M.: A Knowledge Based Vision System for the Automated Adaptation to New Scene Contents, Eighth International Conference on Pattern Recognition, Paris, October 1986.

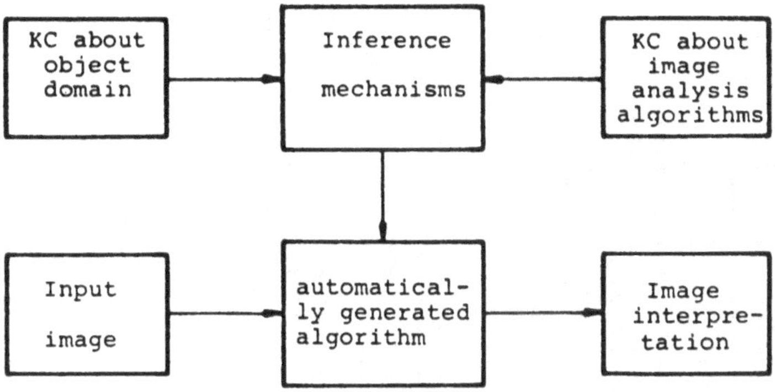

Fig.1. Concept of a fast and flexible image analysis system (KC = knowledge contents)

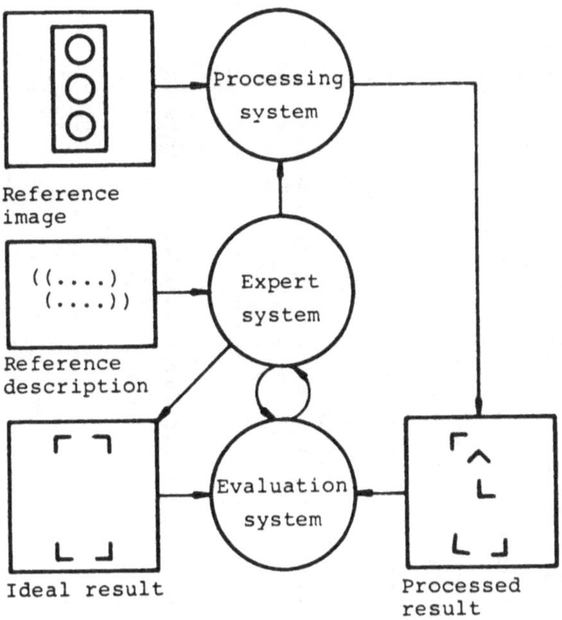

Fig.2. Concept of automated adaptation

Properties \ Coding methods	A L G O R I T H M	P A R A M E T E R	R U L E	S E M A N T I C N E T
Knowledge representation	implicite procedural	implicite declarative	explicite procedural	explicite declarative
Tools for coding of KC	Algorithmic programming languages (FORTRAN, PASCAL, C, LISP, etc.)		Expert systems (OPS-5, etc.)	Data base systems Expert systems
Modification of KC	difficult in general by new formulation of a.	easy by replacement of parameter value	easy by new formulation of rule	easy by modification of data objects
Utilization of KC	by application of algorithm	by reading of parameter value or symbol	by interpretation of the rule and its application	by interpretation using an algorithm or a rule
Efficiency of method	high	high	low	low
Prefered area of application within image analysis	Preprocessing, segmentation, extraction of first symbolic description General: KC which can be formulated by mathematical formalisms		Description of objects and their relations on all levels of abstraction Strategies,plans,hypothyses,inference General: KC which can be formulated by heuristic formalisms	

Fig.3. Coding methods for knowledge contents (KC) and their properties

	Area of application \ Coding methods	ALGORITHM	PARAMETER	RULE	SEMANTIC NET
(a)	Object domain				Relevant object description Reference descr.
(b)	Processing method	Information enhancement			
(c)	Evaluation method	Properties of quality measures	Relevance of scene domain cues	Interpretation of quality value	
(d)	Adaptation strategy			Methodology Evaluation	Rule context
(e)	System control	System shell Communication		Rule selection	Rules Processing proced. Evaluation proced. Activation of procedural components
(f)	General	Scene domain cues		Maximation Minimation	
(g)	Processing algorithm		Optimal parameters		Opt. sequence of processing proced.

Fig.4. Coding of knowledge contents in the automated adaptation system

Quality:

$$Q = 1 + (K-G)/(C*N) + Diff/N$$

$$\text{with } Diff = - \sum_{m \in M} \min_{r \in R} [\, d(m,r)/D \, , 1 \,]$$

Notation used:

R	-	Processed set of scene domain cues
M	-	Ideal set of scene domain cues
N	-	Number of elements of M
G	-	Number of elements of R
K	-	Number of correctly extracted scene domain cues
D	-	Tolerance bound
C	-	See text below
d(,)	-	Weighted Minkowsky Distance

Knowledge contents about properties of the evaluation criterion:

1. Increasing similarity between the ideal and the processed set of scene domain cues increases the numerical value of Q.

2. If n scene domain cues are found exactly in addition to a previous state the value of Q is increased by n/N.

3. A tolerance bound is defined for each scene domain cue. The value of Q does not increase if an additionally detected scene domain cue is on the tolerance bound.

4. The value of Q does not change if for each additionally detected scene domain cue which has been found in its exact position, C scene domain cues have been found outside the tolerance bound additionally, too.

Fig.5. Example for the implicite and procedural representation of knowledge contents about the properties of an evaluation measure. The evaluation measure is used to assess the quality of the adaptation process and is based on the comparison of sets of scene domain cues.

```
• NAME              (GRADIENT_FILTER)
• FRAME-DESCRIPTOR      (ELEMENTARY_PROCEDURE)
• RELATIONS
    ((INPUT             GRAY_LEVEL_IMAGE)
     (OUTPUT            NON-UNIFORMITY_IMAGE)
     (PARAMETERS        G-PARAMETER-1)
     (NEXT_PROCEDURE    LOCAL_RANKING))
```

```
• NAME              (G-PARAMETER-1)
• FRAME-DESCRIPTOR      (PARAMETER)
• RELATIONS
    ((TYPE                WINDOW_SIZE)
     (CURRENT VALUE       7.)
     (DEFAULT VALUE       7.)
     (PERMITTED VALUES    9. 7. 5. 3.))
```

Fig.6. Example for the explicite and declarative representation of knowledge contents about the elementary procedures.

```
• NAME         (RULE-101)
• GROUP        (RULES)
• RELATIONS
    ((CONTEXT ADAPTION-VERTEX-DETECTION)
     (PRIORITY 10.))
---------------------------------------------
• CONDITION
     (FOR_ALL X,Y FOR_WHICH
     (ADAPTION_VERTEX_DETECTION REQUIRES ELEMENTARY_PROCEDURES X)
     (X HAS_PARAMETERS Y))
---------------------------------------------
• ACTION
    ((COPY DEFAULT_VALUE_OF Y TO CURRENT_VALUE_OF Y))
---------------------------------------------
• DESCRIPTION
    IF     1) PARAMETERS FOR THE DETECTION OF VERTICES SHALL BE
              ADAPTED
    THEN   1) SELECT FOR THE CURRENT VALUES OF THE PARAMETERS
              THE DEFAULT VALUES AS INITIAL VALUES
```

Fig.7. Example for the explicite and procedural representation of knowldege contents about the strategy of adaptation.

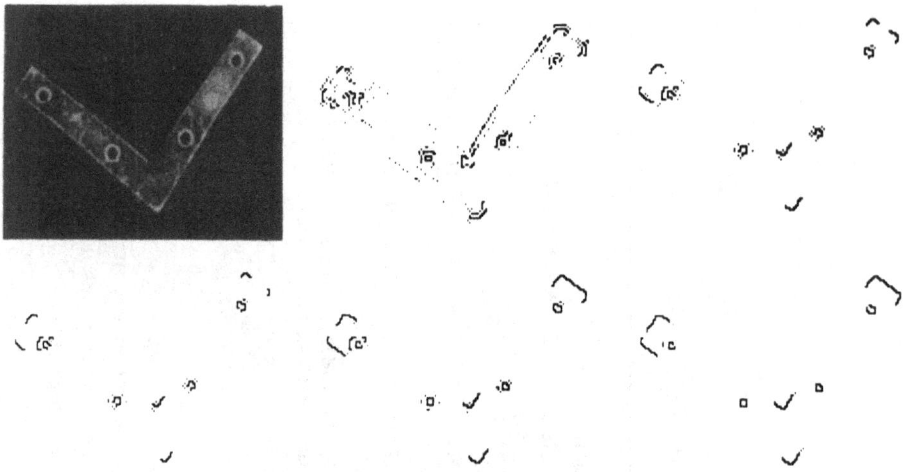

Fig.8. Stepwise adaptation of the processing part for the extraction of vertices.

KNOWLEDGE BASED PROGRAM CONSTRUCTION FOR IMAGE ANALYSIS

R. Hesse, R.Klein, R. Klette 1)

1. Image Analysis and Program Construction

The current trend in computer hardware especially the sixteen bit personal computer development opens a good perspective for small low cost image processing systems. This enables the widespread use of low- and medium-power-systems in different application domains of science and industry. That means that more and more unexperienced and unskilled (according to computer science and programming) domain oriented specialists must be able to operate these systems to solve certain problems of specific application domains. The effective usage of such systems without the necessity of special training requires sophisticated tools providing extended support without the necessity of traditional programming.

The current way to satisfy this requirement is to provide domain oriented program packages to the user society. This causes problems for both the user and the producer of image processing systems. The first must invest a lot of labour in management and documentation, the latter suffers from the insufficient flexibility of those packages which requires programming to adapt them to the specific tasks.

The paper reports the development of a prototype system for knowledge based construction of image analysing programs as an possible approach to overcome the above mentioned difficulties. The system "X...." produces image analysing programs in the source language "...." on the bases of a domain oriented knowledge base provided by the producer of image processing systems for special application domains. The specification of the programs to be generated by the user takes place by interactive problem oriented dialogue. This expert system approach does not open new perspectives of up to now unexplored domains of image processing but helps to transport the existing domain knowledge to the user in an effective and flexible way for both user and producer.

The system follows the principle of hierarchical stepwise refinement of programs. The domain independent forward chaining control

---

1)

Zentralinstitut fuer Kybernetik und Informationsprozesse der
Akademie der Wissenschaften der DDR
Kurstrasse 33, Berlin, 1086, DDR

mechanism of the knowledge interpreter uses a general frame-like representation of control and data flow of sequential programs. This representation includes the description of different levels of program specification during the refinement process.

An application domain oriented knowledge base contains deductive rules and predefined fragments of program structures. After the interactive specification of the program task by means of problem oriented dialogue between operator and system an interchangeable program generator produces the source code of the program in the goal programming language.

Many additional system components support the operator with verbal and graphical help information, provide explanations and overviews about the system's reasoning process, or allow the interaction with the goal system for alternate program specification and program test.

The aquisition of knowledge is supported by a comfortable graphic oriented knowledge base editor which is part of the system allowing the alternation of knowledge modification mode and program construction mode to ensure the consistency and correctness of the knowledge base.

The practical background of the project started in 1985 is to develop a tool which allows comfortable management and flexible usage of application domain oriented software for special PC user groups. Concerning the criterions expense and flexibility from the user's point of view the chosen approach is a compromise between general program languages and one purpose software packages. A first prototype version is applied to the construction of programs analysing medical microscopical slides with a 16-Bit-PC image processing system.

2. Computer aided Program Planning

The specification of computer programs can be considered as hierarchical process of planning actions to achieve a complex goal. The applied principle of stepwise refinement leads to the following method:

(1) top-down refinement of the programming task by stepwise decomposition of complex goals into sets of functional independent but organisational and informational connected subgoals with less complexity until there will be only terminal subgoals which can be achieved by elementary actions (predefined program fragments);

(2) buttom-up construction of the program according to the

goal decomposition of step (1) by arranging predefined program fragments (actions to achieve terminal subgoals) which are connected by predefined data structures to ensure the necessary interaction.

The algorithms of goal specification and program construction are generated by combination of the components of a knowledge base and specified by using information of a data base obtained from the user communication or the input data of the program to be planned. The refinement of one subgoal is independent of other subgoals and considers only local circumstances. The interaction of program actions takes place via global data according to the interaction of the corresponding subgoals.

3. Representation of Program Planning States

A frame concept was developed to represent the actual refinement stage of the program to be planned. Each potential program action is described as special data structure with the following slots:

(1) a verbal description of the goal to be achieved,

(2) functional complexity (terminal goal or not),

(3) successor(s) of the action during program processing,

(4) a pattern for addressing the starting rule of the refinement algorithm,

(5) father and children frame names according to the position within the refinement hierarchy,

(6) a fragment of the program's source code which represents the action to achieve the goal (only in case of terminal goals),

(7) arguments which are data structures and connect the action with others.

The whole program stage is represented by the full list of the frames generated during the past refinement steps. This list represents a so-called program net being a set of related program graphs. Fig. 1 shows a cut-out of an example program net in form of two connected program graphs. The rectangular nodes represent potential program actions (functional nodes) of different organisational type (action, split) connected by successor-relations (organisational program structure or

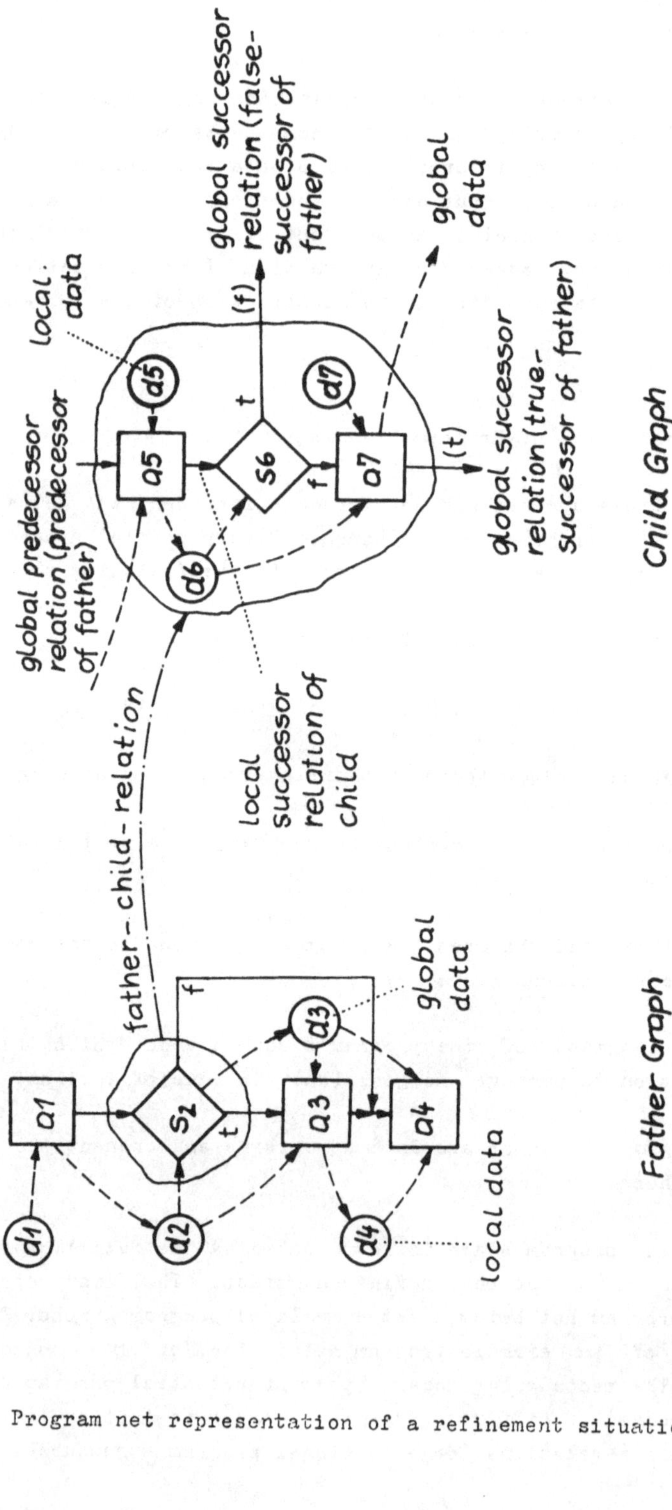

Fig. 1: Program net representation of a refinement situation

control flow). The circular nodes represent data which are connected with functional nodes by input-output-relations to ensure the necessary interaction between the functional completely independent program subfunctions (informational program structure or data flow). The father-child-relation (means that the father node was refined by the child graph) generates a hierarchy of program graphs which layers describe different levels of specification detail of the program function. It also ensures the correct management of global and local data by causing an inhieritage mechanism passing the related data of the father node to the corresponding functional nodes of the child graph.

The program construction process starts with a trivial initial program graph consisting of only one functional node and no data nodes representing the fully unspecified overall program function. The refinement of one goal function node of a certain program graph needs to consider only the connected data nodes and can be processed independently of all other nodes of the whole program net. The complete refinement of the program will be achieved by refining repeatedly all goal nodes until there are only terminal nodes in the lowest levels of the program graph hierarchy.

3. Knowledge based Refinement

The basic algorithm of goal node decomposition and action generation is as follows:

(1) characterization of the goal node function to be decomposed during user-system-interaction and storing the obtained predicate values in the data base,

(2) classification of the goal node function applying production rules of the knowledge base which match the given goal node function characteristics of (1),

(3) allocation of a predefined subgoal structure belonging to the infered goal node class (childgraph of the goal node),

(4) specification of the subgoal structure by specification of its data structures in order to fit it according to the real conditions of the goal node to be refined.

This algorithm is applied repeatedly to each goal node until there are only terminal nodes within the frame list. Then a program generator will produce the source code of the program. Therefore the knowledge base contains two different kinds of knowledge units: production rules and predefined subgoal structures.

The units of the knowledge base are addressed by formal string patterns. Because of the fact that slot (4) of the frame representation of functional nodes may contain such an address pattern as a pointer to the rule first to apply in case of refinement of the node, the elements of the knowledge base are ordered within a generalisation-specialisation-hierarchy. This hierarchy, the introduced substructures, and rules determine the domain of application of the system. All knowledge is declared with the help of a frame-like description language and can easy be changed or modified.

To prepare the system for a special application domain the knowledge can be inserted with the help of a comfortable graphic-supported knowledge aquisition module.

The separate program generator allows easily to adapt the system to different goal programming languages. Fig. 2 shows a flow chart of the system kernel.

4. Application to Image Processing

The system was implemented in PASCAL and runs under MS-DOS operating system on IBM-PC-compatible 16-Bit-PC's with hard disk external memory and 640 Kbyte operational memory.

Firstly it was applied to develop programs for the analysis of medical microscopical slides. The knowledge base contains about hundred

(1) basic program structures for

- repeatedly processing of series of pictures
- processing of single pictures

(2) substructures for

- picture input
- picture preprocessing
- object search and object isolation
- object classification
- object class evaluation
- result output

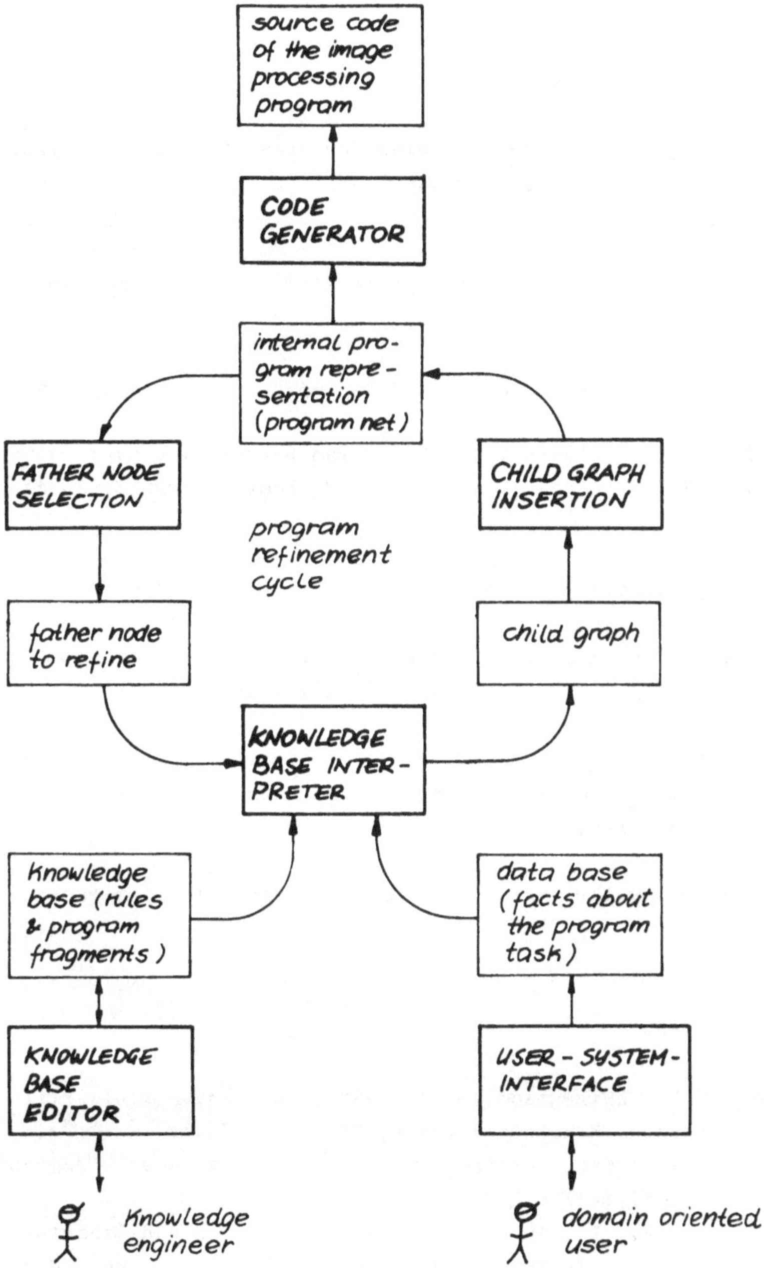

Fig. 2: Flow chart of the system's kernel

(3) data structures for
   - class membership criterions
   - class features
   - output formats
   - parameters of functions, and

(4) simple menue driven rules for specification of program tasks and choosing functional substructures from the knowledge base.

The generated picture processing programs are written in the language LAMBA (Robotron,1985).

During the test phase the following results could be stated:

- The maximum time-out of the system between two user requests is some seconds. The development of a typical program takes about half an hour.

- The generated programs are well structured and commented.

- The efficiency of the automatically generated programs doesn't sicnificantly differ from that of hand-made programs.

- Additionally inserted verbal or graphic help and overview functions make the handling of the system very comfortable to the domain-oriented user.

Up to now the system was tested in analysis of microscopical slides in pathology.

5. References

Cohen, P. R., Feigenbaum, E. A., 1982, The Handbook of Artificial Intelligence, Heuristech Press, Stanford, C. A.

Robotron Vertrieb Berlin (VEB), 1985, Dialog-und-Programmier-System AMBA/R, Berlin

Sacerdoti, E. D., 1977, A Structure for Plans and Behavior, Artificial Intelligence Series, Elsevier North-Holland Inc., New York

Voss, K., Hufnagl, P., Klette, R., 1985, Interactive Software Systems for Computer Vision, Progress in Pattern Recognition 2 (Kanal, L. N., Rosenfeld, A., eds.), North Holland, 57:78

Stefik, M., 1981, Planning with constraints (MOLGEN: Part 1 and Part 2), AI-Journal 16 (1981), pp. 111-170

AN ASSOCIATIVE NETWORK AS SYSTEM SHELL FOR KNOWLEDGE BASED IMAGE UNDERSTANDING *

G. Sagerer, S. Schröder, H. Niemann
Lehrstuhl für Informatik 5 (Mustererkennung)
Martensstr. 3
D-8520 Erlangen
Federal Republic of Germany

Abstract:
For the analysis of complex images, or image sequences explicit knowledge about the task domain is required, in order to extract a symbolic description out of the input data automatically. In this paper, a shell for such image analysis systems is described. It combines a powerful knowledge representation scheme with a control algorithm, which may be refined for special tasks. Tools for knowledge acquisition complete the shell. It does not depend on a special field of problems to be attacked. It was successfully used for the analysis of scintigraphic image sequences of the heart. Applications for the analysis of industrial scenes, and speech understanding are in progress.

1. Introduction

The purpose of image analysis is the automatic investigation of a symbolic description of an image, or an image sequence. The content of the symbolic description must correspond to the requirements of a concrete application. E.g., for a medical application like X-ray images, diagnostic descriptions are to be extracted out of an image; for industrial scenes descriptions of the quality of built objects or special tasks for a robot are required. To achieve this, complex applications need the explicit representation of extensiv knowledge about objects to be recognized, their relations to each other and their task specific domain. Furthermore there is a need for algorithms to use the stored knowledge during the analysis process. The efficiency of an image analysis system therefore depends on both the knowledge base and the control algorithm. Besides this two modules, knowledge base and control, a database for the results and methods for preprocessing and segmentation of images are necessary/1/. In many approaches the system is completed by an explanation module which allows the user to check system activities and to evaluate final and intermediate results after an analysis process has been finished. Up to this point the structure of an image analysis system, compare Fig.1, is quite similar to a so called expert system/2/. The differences are mainly the input data and therefore also the data transformation processes. While expert systems transform symbolic data into other symbolic data, image analysis systems has to extract symbolic descriptions out of image matrices, additionally. This yield in a second distinction. In expert systems input data can be viewed as to be correct, while images are noisy, segmentation errors may occur, and so on. That is the reason why image analysis systems have to take into account certainty values of data they manipulate. Viewing the knowledge base of a system, different approaches for formalizing knowledge exist. If a special approach is fixed, knowledge acquisition is done by hand and the obtained knowledge is coded into the desired approach. First investigations are done in the automatic acquisition mainly in rule based approaches. Up to now, a rough knowledge is also necessary for training processes and training is only possible for a fixed representation scheme. The acquisition modules, basic knowledge and training processes completes the system structure in Fig. 1.
Depending on the scheme used for knowledge representation, image analysis or more general pattern analysis systems can be divided into the following groups:

- syntactical approaches based on formal grammars
- relational database systems
- rule based systems
- logic systems based on first order predicate calculus
- associative network systems

Overviews and descriptions of different systems are given for example in /1,3,4/. Control algorithms mainly used are meta rules, depth first or best first search, and the A^*-Algorithm/9/. To handle certainties heuristic functions, probabilities, or fuzzy sets are used.
In this paper a system shell for pattern analysis systems will be described which depends on an associative network formalism and the A^*-Algorithm. In this shell, an

* The work reported in this paper is partially supported by SIEMENS AG, Erlangen

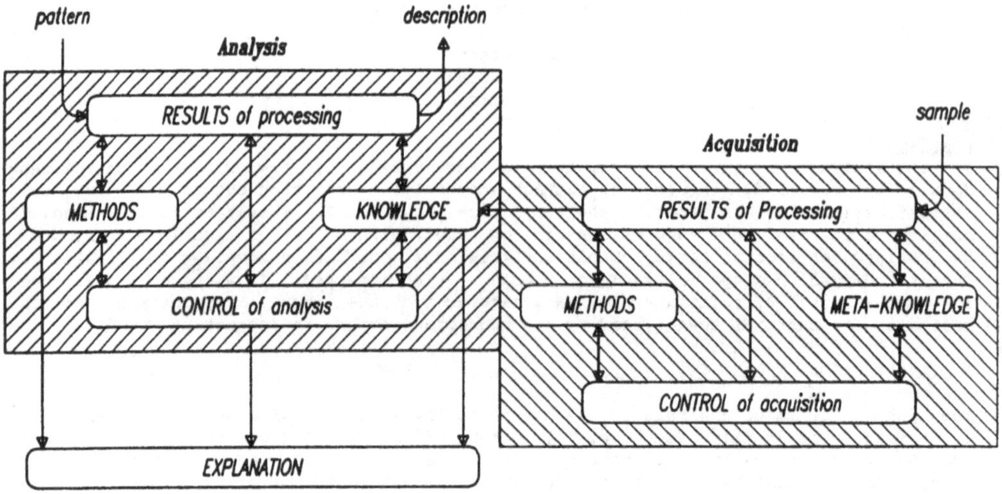

Fig. 1 Modules of a Pattern Analysis System and their Interactions

active hierarchical knowledge base incorperates with an control algorithm, which only knows the syntax and semantics of the network but not it's content. This allows, to create homogenuous knowledge based systems for various applications. Therefore, theshell itself is independend of a special application. Besides the knowledge base structure and the control module which may be refined, the shell actually subsumes also tools for interactive acquisition, for creating and updating the database for results and allows different types of certainty-functions, only restricted by the A^*-Algorithm. Tools for at most automatic knowledge acquisition are in development. The shell was successfully used in a system which achieves automatically an diagnostic interpretation of scintigraphic image sequences of the human heart. Two further systems, a speaker independent continuous speech understanding system and for the interpretation of industrial scenes are in progress.

2. Knowledge Representation by an Associative Network Approach

2.1 General Remarks

Associative (or semantic) networks are special directed graphs, with labeled nodes and edges. First developed to represent the semantics of natural language, they were used as knowledge representation scheme for different applications. The basic idea is quite simple: Information about conceptions (objects, relations, events) are represented by nodes, interrelationships between nodes are represented by labeled edges, to mark also the kind of the relationship. Additional, a special kind of edge, mostly named "IS_A", is used to connect a conception to another one which represent a class. Along such a hierarchie of more general to more special nodes an inheritance of the properties is defined. The theoretical background of such networks was studied with large effort by many groups/5/. It was carried out, that associative network formalisms has to be epistemological adequate. That means on the one hand that there is a need for a clear distingtion, what kind of information has to be represented by nodes and what kind by edges/6/. On the other hand the syntax of a network must be well defined and also the semantics of the different types of nodes and edges must be fixed/7/. This leads to formalisms which are independend from the special content of knowledge they are used for. Further investigations were done to expand the prior only declarative content of such networks due to procedural knowledge. Starting with procedural attachment at some slots of a node this yield to the powerful integration of procedures into "procedural associative networks", e.g./8/. By our opinion, such well structured networks with the capability to handle also procedural knowledge are, at the moment, the adequate way to built up knowledge bases for image (and speech) analysis systems. Our approach which will be described in the next sections is mainly influenced by KL-ONE/7/ and PSN/8/.

2.2 Types of Nodes

Three different types of nodes exist in the network. First of all, a *CONCEPT* gives the intensional description of an object, an event, or classes of objects or events. The intensional description subsumes all properties of the conception to be modeled. Examples of such concepts are general ones like "3D-object", "movement", or "disease", or more special ones like "car", "run", or "enlarged left ventricle". The second type, called *INSTANCE*, represent a hypotheses that a real world object or event satisfies the description of a concept. If, for example, some region of an image is interpreted to be a "car" during the analysis process, an instance of car is created referring this region, may it be true or not. Therefore one real object may be referred to by more than one instance, but each instance is associated with exactly one concept. The concepts form the knowledge base, while their extensional sets, the instances, are the intermediate and final results. Viewed as data structures, they are both identical, exept that instances hold results while concepts subsume descriptions. The third type of node, the *META-CONCEPT*, build up the knowledge base for the acquisition process, which results in a knowledge base for the image analysis system. Concepts can be looked at to be created by meta-concepts and data. This third kind of node will be described in section 4. The features of concepts and their interrelation will be discussed in the following. Fig. 2 shows the syntactical structure of a concept. Hatched blocks are necessary for the knowledge acquisition process and are explaned in section 4.

2.3 Types of Edges

Relationships between concepts (and also between instances) can be built up by three different types of edges. They are denoted by *generalization*, *semantic-part*, and *necessary-part*. Each of these edges is described in a concept, see Fig.2. While a concept can only have an unique generalization which is another concept, it may consist of arbatrary number of semantic-, and/or necessary-parts. All the edges define a hierarchie in the network. Both part relationships are dscribed by a complex slot, see Fig.2. If, e.g., a concept A has concept B as semantic-part, B is the *domain* inside a slot semantic-part in A. For all the three relationships also the inverse is defined by *specialization*, *semantic-part-of*, and *necessary-part-of*, respectively. In the following, first the semantics of the relationships will be discussed, and after that the different facets inside the both "part"-slots. *Generalization* combines a concept to another one which represent a superclass compared to the concept itself, e.g. vehicle would be the generalization of car and object of vehicle. Like in KL-ONE all properties, that are parts, attributes, and structural_relations defined in a concept A are inherited to all concepts which are specializations of A, but they may be explicitely modified. Such modifications are to be described in special facets of the different slots. The restrictions and the rules for such modifications are comparable to those defined in KL-ONE. Contrary to other approaches of semantic networks we distinguish two different "part" relationships. *Semantic-parts* of a concept must hold the restriction that they are parts in a natural sense. E.g., semantic-parts of a car are wheels, semantic-parts of a special movement can only be other special movements, not an object. Therefore we can denote this by: a concept and its semantic-parts must have the same degree of abstraction. By the relationship *necessary-part* concepts of different abstraction levels can be combined. If a concept A has necessary-part B, this fact can be expressed by: to talk about A requires B. Therefore a object can be necessary-part-of a movement, or a line of a car. All slots in a concept are referred uniquely defined by their *role*, which describes the functional role of the slot inside the concept. E.g., the fact that one car passes another one has to be modeled in the following way: a concept "pass" has two necessary-parts. Both have the domain "car". One of the part-slots has the role "passes", the other one "is passed". The facet *modified* explain modifications of an inherited slot. The value can be NO, if the slot is not defined in a more general concept, YES, if a slot with identical role has been modified compared to the more general concept, and therefore been overlayed by a new definition. If also the role has changed, the old one has to referred to by this facet. Finally, the facet *computation*, which is only defined for semantic-parts (and attributes) gives information about the context dependency of the referred part. That is, the conception defined by this part depends on a larger context, i.e., the concept referring the part. E.g., the conception "(chair-)leg" depend on the context "chair". It is not possible to describe a concept "chairleg" without the knowledge that it is a part of a "chair". Contrary, a wheel can be modeled without knowing if it is part of a car, or a bike and so on. Different semantic-parts or different necessary-parts can be combined to *sets of modality*. Each such set describe which combinations of such parts are sufficient for the concept (*obligatory set*), and for each set the *optional* and *inherent* parts which may complete the optional set.

concept

slot	value
name of concept	text
hierarchy levels	4 int
priority	6 int
descriptive info.	text
generalization	**concept**
spezialization	list of **concepts**
context	list of **concepts**
meta-concept	**concept**
model-concept	list of **concepts**
semantic-part-of	list of **concepts**
semantic-parts	list of **parts**
sem.-attributes	list of **attributes**
sem.-structures	list of **structures**
necessary-part-of	list of **concepts**
necessary-parts	list of **parts**
nec.-attributes	list of **attributes**
nec.-structures	list of **structures**
local-attributes	list of **attributes**
modality	list of **modality-sets**
CF-procedure	name of procedure
CF-arguments	list of arguments
argument-test	name of procedure
graphic	X- and Y-value

modality-set

slot	value
obligatory	list of **parts**
optional	list of **parts**
inherent	list of **parts**

default-value

slot	value
typ	int or float or char
number	int
values	(typ)-array

attribute

slot	value
role	text
differentiation	2 int
comp. of diff.	name of procedure
dimension	2 int
comp. of dim.	name of procedure
computation	name of procedure
arguments	list of arguments
argument-test	name of procedure
inverse comp.	name of procedure
restrictions	name of procedure
domain	int or float or char
selection	2 int or 2 float
modified	YES or NO or role
mmodified	NO or role
default-values	**default-value**
comp. of default	name of procedure

part

slot	value
role	text
differentiation	2 int
comp. of diff.	name of procedure
dimension	2 int
comp. of dim.	name of procedure
domain	name of **concept**
computation	YES or NO
restrictions	name of procedure
modified	YES or NO or role
mmodified	NO or role

structure

slot	value
role	text
differentiation	2 int
comp. of diff.	name of procedure
relation	name of procedure
arguments	list of arguments
argument-test	name of procedure
default-value	float
mmodified	NO or role

Fig. 2: Relationships, Slots, and Facets building up the definition of a Concept

2.4 Attributes, Structures, Certainty

Attributes characterize properties of a concept which are essential with respect to the task domain. In most cases such attributes are physical properties, like the volume or the color of an object, or start or end time of a movement. But also, e.g., textual explanations of diagnostic interpretations may be attributes of concepts. But they must not be used to handle properties which describe relationships to another attribute or special relationships between properties of parts. Identical to parts, attributes are uniquely defined by their *role*. Additional facets give the *domain*, *selections* for the domain and *default* (or initial) *values*. As mentioned above, attributes are inherited from a more general to a more special concept. But it is also possible to *modify* an attribute in the same way as a modification can be done for parts. Substantial for the definition of an attribute is the procedure how a value for an instance can be calculated (*computation*). Such a procedure can be a simple multiplication, but also a parser combined with a grammar, or, even more complex, a complete rule based system. There is no necessity for such procedures to give an unique result. Based on the same input data they may build up different competing results. Each is stored in one instance, and therefore competing instances are created, each having one unique value for all its attributes. The results of the computation can be restricted due to the conception they are used for. The procedure referred to by *restrictions* do this job. The *arguments* of the computation procedure are referred in a further facet of the slot. The arguments are a subset of those attributes which are defined in parts of the concept, or if the concept depend on a larger concept also in the concept describing this context, compare section 3. In order to support control algorithms with mixed bottom-up and top-down strategies, also a procedure for an *inverse computation* is referred.
Properties which must hold between different attributes and/or attributes of parts, are to be tested by *structures*. The semantics of these slots, which are also uniquely defined by their *role*, is: a *relation* tests the *arguments* due to the fact substantial for the concept. E.g., in a concept "car" a structure "above" may be defined, which is able to test, whether the "bodywork" is above the "wheels".
The results of structural tests are essential for an instance to be more or less valid for the concept. Other informations which may help to judge the quality of an instance are the restrictions facets in the different slots and the quality of the instances referred to by part relationships. Out of these facts the *CF-arguments* can be chosen to check the quality of an instance by the *CF-procedure*.

3. Use of Knowledge, Creating Instances

The main idea to use knowledge stored in an associative network as described above is concentrated in the *basic rule for instantiation*. This rule completes the definition of the network scheme by a third step. Besides the syntax and the semantics of the formalism, the use of stored knowledge and therefore the pragmatic is defined. This can also be done without respect to a task domain. The rule can be expressed as follows:

IF for a concept A with respect to one *obligatory-set* of the *modality* of A, instances for those concepts exist, which are referred to by the following slots in A or slots inherited to A without *modifications*:
  necessary-parts AND
  semantic_parts with facet *computation* equal to *NO* AND
  one of *context* if this slot is not equal *NO*

THEN build up instances ivp(A) of A as follows:
- create empty instances of A,
- connect these instances to those instances, which are mentioned in the condition of the rule,
- calculate the *attribute*-values for all ivp(A), with respect to the *arguments*
- test the *restrictions*
- estimate the quality of ivp(A), by *CF-procedure* with respect to the *CF-arguments* referring yet known instances and *restrictions*

By this rule partial instances are built up with calculated attribute values and tested restrictions, and an estimated quality. As mentioned in section 2.4, for structural tests attributes out of all parts may be used. Therefore an additional rule, the *rule for completing an instance*, is required:

IF an instance ivp(A) of a concept A exists AND instances for all those concepts exist, which are referred to by the *semantic-part* slots in A, with respect to the *obligatory set* of the *modality* used for building up ivp(A)

THEN build up iv(A) out of ivp(A) as follows:
- connect iv(A) to the instances referred to by the condition of this rule and not yet connected to ivp(A)
- test the ***structures*** defined in A with respect to the ***arguments***
- calculate the quality of iv(A), by ***CF-procedure*** with respect to the ***CF-arguments***

If a goal concept for an analysis process is known, recursive application of these both rules results in a search path for the goal concept. By competing instances, generated for a concept, this path is splitted into a search tree. Based on the qualities of instances qualities for concepts restricted to one path of the tree can be estimated. This yield to quality functions for the nodes of the search tree, which are used for the A^*-Algorithm /10/. Up to this point only obligatory sets of the modality of the concepts were used. An additional rule, the ***rule for extending an instance***, also handles the optional and inherent parts.

IF an instance iv(A) of a concept A exists AND instances of concepts exist, which are ***optional*** or ***inherent*** due to the ***modality*** used for building up iv(A)

THEN extend iv(A) to ive(A) as follows:
- connect the instances mentioned in the condition to ive(A)
- calculate ***attributes*** with restect to the ***arguments***
- test ***structures*** with respect to the ***arguments***
- calculate the ***CF*** of the new ive(A)

If the goal concept is known, this three rules together with the A^*-Algorithm form the skeleton for different control strategies. But they are also one complete strategie. Furthermore, it is also possible to extract automatically potential goal concepts /15/.

4. Knowledge Acquisistion

4.1 General Remarks

The ability of knowledge acquisition (learning) is very important for modern image analysis systems. Although the analysis system is designed modularly, the adaption to another field of interest is a time spending process. The bottleneck of this process is of course the construction of the new knowledge base.
There are basically two approaches to knowledge acquisition: leave it to the designer, that is, the manual approach, or leave it to the machine, that is, the (automatic) learning approach. And there is an arbitrary number of related approaches resulting from some mixture of the manual and the learning approach /1/. Examples of knowledge acquisition for image analysis systems are given in /13,14/.
Until now we followed the manual approach. The system designer was able to build up the declarative part of the knowledge base using a special network editor. This editor is described in /12/. The goal of our knowledge acquisition system will be an almost automatically construction of the declarative part of the knowledge base. This learning will be done by examples. Our knowledge acquisition system will enable persons, who are not well versed in acquisition networks, to build up a new knowledge base. This is an important precondition for an easy adaption of the analysis system to different fields of interest.

4.2 Features used by the Acquisition Process

The hatched blocks in Fig. 2 show the features which are used by the acquisition process. If a concept is generated by the acquisition process using a special meta-concept, these two concepts are connected by an edge named ***meta-concept*** from the generated concept to the meta-concept. The inverse edge is named ***model-concept***. A concept of the knowledge-base of the analysis process can only have one meta-concept while a meta-concept can have several model-concepts. There is no inheritance from a meta-concept to its model-concepts. Instead each meta-concept contains a description about how to construct such a model-concept.
During the acquisition process the existing parts, structures, and attributes of the meta-concept may be split. For instance the meta-concept 3D-OBJECT may have a part surface. Of course the generated concepts of real objects will have several surfaces. Therefore the process of splitting, named differentiation, is controled by a given interval. This interval defines the allowed number of differentiations. The slot ***differentiation*** contains the bounds of this interval. The slot ***computation of differentiation*** refers to the procedure that computes the real number of differentiations. The same problem occures in the case of dimension. Different model-concepts of the same meta-concept may have different dimensions of their parts, structures and attributes. The problem is solved in the same way as the differentiation. The slot ***dimension*** con-

tains the bounds of the allowed dimension-interval while the slot ***computation of dimension*** refers to the procedure which computes the real number. The slot ***mmodified*** is used if a part, structure or attribute of the meta-concept has to be splitted. In this case the slot of the model-concept contains the role of the part, structure or attribute of the meta-concept. Otherwise ***mmodified*** is set to NO.
The slot ***argument-test*** always refers to a procedure that checks the existance of arguments of procedures that are used by the analysis process.
The slot ***computation of default*** refers to a procedure that computes the default-values during the acquisition process.

4.3 System Structure

The system structure of the acquisition system is shown in Fig. 1. It is similar to the structure of the analysis system. The modul METHODS is equal to the same modul of the analysis system. The content of the modul META-KNOWLEDGE will be explained later. During the acquisition process the database for the results contains concepts instead of instances in the case of analysis. The control of the acquisition process is done by a special algorithm which only knows the syntax and semantic of the assosiative network but not it's content. Of course this algorithm is not as sophisticated as the control algorithm of the analysis process. The acquisition process does not have to classify the input data. The goal is to determine a symbolic description of the input data of a given object or a given class of objects that can be used by the analysis process. In our system the symbolic description is stored in an associative network. The two steps of the acquisition algorithm are: creating one concept using a single observation and creating a new concept by comparison of existing concepts. The input data is a sample. But this sample may not only consist of images. For example the process for acquisition of workpieces may use images and CAD data of these workpieces as input data.

4.4 Meta-Knowledge

The knowledge base of the acquisition process contains the meta-knowledge. It is separated into a declarative and a procedural part. The declarative part is the meta-model which is formed by the meta-concepts. The meta-model is based on the associative network formalism, which was explained in chapter 2.
The procedural part of the meta-knowledge consists of the procedures referred to by the slots ***argument-test***, ***computation of differentiation***, ***computation of dimension***, and ***computation of default*** and of the rules used to compare intermediate results.
The meta-knowledge incorperates with the control algorithm of the acquisition process. But of course they are independent in the way that the same algorithm will work with different meta-knowledge each for one field of interest.

4.5 The Acquisition Process

The acquisition process is divided into two steps. The first one is to create one concept using a single observation. This observation may be for instance an image of an object or CAD-data. The algorithm has two parameters. The first parameter is the name of the meta-concept that leads the algorithm. The second one is the name of the concept to be created.
During the first phase of step one the acquisition algorithm will build up a ***concept-shell***. ***Concept-shell*** means that the new concept, it's parts, attributes and structures, and the edges between the parts and the concept are created, but the slots which will later contain special values are empty. This means during this phase computation of differentiation and computation of dimension takes place. The parts, attributes and structures of the meta-concept may be splitted.
The goal of the second phase is to test the existance of the arguments of the analysis procedures. This is done by calling the routines referred to by the slots ***argument-test***. A part, attribute or structure of the meta-concept may no more occur in the model-concept because of differentiation. If a part, attribute or structure of the meta-concept is splitted into several parts, attributes or structures, then they will have new and different roles and they can not have the role of the part, attribute or structure from the meta-concept. In this case the names of the arguments which are composed of the roles have to be determined again.
During the third phase, the default values are computed by calling the computation procedure of each attribute. The analysis process may be improved by restriction of the domain of attributes. But if, for example, the analysis is supposed to recognize an object independent from its position and the computation procedure for a special 3d-point computes the position values, it would be fatal to restrict the domain of this attribute. On the other hand it may be useful to restrict the domain of an attribute to

an interval close to the computed default value. This way the analysis will only search for segments of the image that satisfy the given restriction. The conclusion from these examples is: Restriction of the domain is not always useful and if a restriction takes place, some inaccuracy should be taken into account.
The concepts created by the first step are the intermediate results of the acquisition process that are stored in the database. The second step will create new concepts by rulebased comparison of these intermediate results. The comparison may be done to combine several observations of the same object. Especially if you want to build up a complete 3d-model of an object you have to examine several images from different views. On the other hand the comparison of concepts may lead to a description of a given class of objects (generalization). This can be done by specifying positive and negative examples.
After these two steps have been processed several times the final results of the acquisition process, the created concepts, have to be transmitted to the knowledge base of the analysis process.

5. Applications and Results

The shell described was used for the automatic diagnostic interpretation of scintigraphic image sequences of the heart. This system is discussed in detail in /10/. The knowledge base consists of about 180 concepts with more than 1000 edges. Procedures to calculate attributes are, beside others, also a parser with formal grammars, dynamic programming routines, and fuzzy rules. All the tests and qualities are realized by fuzzy-membership-functions. The system was tested with 21 image sequences. 18 were interpreted correctly, while 3 had result "undefined". A speech understanding system using this shell is in progress. A rapid prototyted version with about 200 concepts is tested. The strategie of this system is mixed bottom-up and top-down due to the basic rule of instantiation. Therefore also "inverted" attribute calculation is used. Datadriven the knowledge base is restricted to created instances. These variable knowledge bases are build up by *MODIFIED-CONCEPTS*. The quality of instances is based on probabilistic features. The main point of the third application, interpretation of industrial scenes, will be automatic knowledge acquisition. The meta-knowledge base of this system consists at the moment of about 20 meta-concepts.
The complete shell, including further investigations, is, respectivelly will be realized in C under UNIX. The reason is the high efficiency. Other versions with less functionality were realized in FORTRAN /10/ and LISP /15/.

References:

/1/ Niemann, H.: Pattern Analysis. Berlin, Heidelberg, New York: Springer 1981

/2/ Hayes-Roth, F., Waterman, D.A., Lenat, D.B. (eds): Building Expert Systems. McGraw Hill 1982

/3/ Ballard, D.H., Brown, C.M.: Computer Vision. Englewood Cliffs: Prentice-Hall 1982

/4/ Hanson, A.R., Riseman, E.M. (eds): Computer Vision Systems. New York: Academic Press 1978

/5/ Findler, N.V. (ed): Associative Networks. New York: Academic Press 1979

/6/ Woods, W.: What's in a Link: Foundations for Semantic Networks. In Bobrow, D., Collins, A. (eds): Representation and Understanding. New York: Academic Press 1975

/7/ Brachman, R.J.: On the Epistemological Status of Semantic Networks. In /5/, 3-50.

/8/ Levesque, H., Mylopoulos, J.; A Procedurel Semantics for Semantic Networks. In /5/, 93-121.

/9/ Nilsson, N.J.: Principles of Artificial Intelligence. Berlin, Heidelberg, New York: Springer 1982

/10/ Sagerer, G.: Darstellung und Nutzung von Expertemwissen für ein Bildanalsesystem. Berlin, Heidelberg, New York: Springer 1985

/11/ Niemann, H., Sagerer, G: Semantische Netze als Ansatz zur Repräsentation und Nutzung von Wissen für die automatische Bildanalyse. In Robotersysteme 1, 1985, 139-150.

/12/ Hofmann, I., Gämlich, R., Niemann, H.: A Human Interface for Control of an Image Processing System. Proc. 8th ICPR, Paris, 1986, 1256-1258.

/13/ Perkins, W.A.: INSPECTOR: A Computer Vision System that Learns to Inspect Parts. IEEE PAMI, Vol.5, No.6, 1986, 584-592.

/14/ Yashida, M., Isuji, S.: A Variable Machine Vision System for Complex Industrial Parts. IEEE Trans. on Comp., Vol.C-26, No.9, 1977, 882-894.

/15/ Eichhorn, W., Niemann, H.: A Bidirectional Control Strategy in a Hierarchical Knowledge Structure. Proc. 8th ICPR, Paris, 1986, 181-183.

THE THREE BASIC FORMULAS OF NEIGHBORHOOD STRUCTURES

Reinhard Klette [1)] and Klaus Voss [2)]

Abstract. It will be shown that the topological theory of neighborhood structures may be based upon the Vertex Theorem, the Mesh Theorem, and the EULER Theorem. These three formulas will be characterized to be of basic topological importance by deducing theoretical facts of practical importance to image processing, basic algorithms, and measures and features of digital point sets which may be considered to be "object areas". The notion of a mesh is of central importace to the presented approach and characterizes a complete period of an oriented way in a neighborhood structure. The orientation is defined by cyclic enumeration of all the point neighborhoods.

1. Introduction

The theoretic-mathematical foundation of a scientific theory is connected to two fundamental preconditions,

- the discipline (notions, methods, results) has to be well-developed itself, and
- by the different branches of Mathematics the tools necessary for formulating the theory have to be available.

During the last two decades in the field of Digital Image Processing several more or less successful attempts may be recorded to formulate a self-contained theoretic-mathematical approach covering topics as connectivity, digital topology, or digital geometry; compare [1, ..., 10] - just to cite a few of related books and papers. In [12] a relatively closed representation of topics mostly related to "low level vision" is given essentially based upon the theory of random point sets [11]. Since 1985 we have published a series of papers [13] which is characterized by the use of irreflexive and symmetric neighborhood relations in discrete point sets as the basic objects of an "image processing theory". Some of the most important facts contained in this series of papers will be presented in this paper in a brief way.

The subjects of digital image processing are images. For dealing with concrete images of a given field of application a certain model has to be selected specifying assumptions on digital topology or geometry. These models were generated by abstraction, and there are some "favorite models" characterized by the so-called "4-, 6-, or 8-neighborhood", for example. On the other hand, such models may be obtained by interpretation of some variables of a given theory which should be a generalization of all the con-

1) Zentralinstitut für Kybernetik und Informationsprozesse, Kurstraße 33, Berlin, DDR-1086

2) Friedrich-Schiller-Universität, Sektion Technologie, Universitätshochhaus, Jena, DDR-6900

crete models relevant to image processing. At this level of abstraction, variables of the theory are points, a certain neighborhood relation defined on the set of these points, and a certain orientation defined by cyclic ordering of all the finite neighborhoods of points. By interpreting these variables, for example points by points of an orthogonal, or by cells of a cellular space, or by regions of images etc., concrete models may beobtained. At this level of a fundamental theory, some established branches of Mathematics may be characterized to be the methodic tools, as Set Theory, Graph Theory, Theory of Algebraic Complexes [14], or Grid Theory [15]. But sets, graphs, complexes, or grids are not the primary subjects of a theory of digital image processing. This point of view may be compared with the investigation of spatial motions in Physics, for example: The subject is a model of the moving mass point, and the theories of vector spaces and differential equations are providing the tools.

So far we see the following four main topics of a theory of image processing which may be studied in the same methodical line,

- the topological theory of neighborhood structures with specific models as support structures of digital images,
- the geometric theory of digital figures on neighborhood structures,
- the metric theory of figures in regular grids based upon mappings of neighborhood structures into the real space, and
- the theory of discrete image functions defined on regular grids characterized by a principal exclusion of continuosity.

In the present paper we shall restrict ourself-s to the most fundamental theory, the topological theory of neighborhood structures.

2. Neighborhood Structures

A neighborhood structure is defined to be a tuple (P, N) where P is a finite set of points, and $N \subseteq P \times P$ is an irreflexive and symmetric relation called neighborhood relation. By interpretation of P and N we obtain concrete models of neighborhood structures. A neighborhood structure (P, N) is isomorphic to an undirected graph without elementary loops and multiple edges. The set $N(p) = \{q \mid q \in P \;\&\; (p, q) \in N\}$ is the neighborhood of point $p \in P$, and $\nu(p) = \text{card}(N(p))$ denotes its valence. Let $\varepsilon = \text{card}(P)$ be the number of points, and let $\varkappa = \text{card}(N)/2$ be the number of edges of a given neighborhood structure (P, N). Then, a basic formula of graph theory is given by the Vertex Theorem

$$\sum_{p \in P} \nu(p) = 2\varkappa . \tag{1}$$

Then, $\bar{\nu} = 2\varkappa/\varepsilon$ may be called the averaged valence of the given neighborhood structure. By graph theory some common notations are given: A way $(q_1, \ldots, q_n)$ in a subset $Q \subseteq P$ is an ordered sequence of points $q_i \in Q$ satisfying $q_i \in N(q_{i-1})$, for $i = 2, \ldots, n$. A set $Q \subseteq P$ is connected if and only if (iff) there exists a way $(p, \ldots, q)$ in Q, for any pair of points $p, q \in Q$. A *region* is a connected subset of P, and a *component* of a set $Q \subseteq P$ denotes a maximal region $R \subseteq Q$.

3. Oriented Neighborhood Structures

By cyclic ordering of all points of each of the neighborhoods N(p), an *orientation* may be attached to a given neighborhood structure. For example, the direction code [1] commonly used in image processing is a specific realization of such a cyclic ordering. A cyclic ordering $\langle q_1, q_2, \ldots, q_n \rangle$ of all the points $q_i \in N(p)$ is called a *neighborhood cycle* c(p). Let C be the set of all neighborhood cycles c(p), $p \in P$. Then, $\Gamma = (P, N, C)$ is the formal representation of an *oriented neigborhood structure*, briefly called *structure* in the sequel.

A (directed) edge $(p, q) \in N$ is *followed by* a (directed) edge $(q, r) \in N$ iff $c(q) = \langle \ldots, p, r, \ldots \rangle$, i. e. point p is a direct predecessor of point r within the neighborhood cycle c(q) of point q. Thus, any (directed) edge (p, q) uniquely defines an (directed) edge (q, r) as its direct successor within the structure (P, N, C). In this way, any edge $(p, q) \in N$ uniquely *generates* an infinite sequence (p, q), (q, r), (r, s), ... of edges, and $w(p, q) = (p, q, r, s, \ldots)$ denotes the *way* of points passed through by this generated sequence of edges. Because the set P of points is finite, any generated way (p, q, r, s, ...) is cyclic. A complete period of a generated way w(p, q) is called a *mesh*, and denoted by m(p, q). This basic notion of our approach is illustrated by Figure 1. The neighborhood cycles coincide with the clockwise ordering of edges, e. g. by this agreement $c(p) = \langle r, t, q \rangle$ or $c(q) = \langle r, s, t, p \rangle$. The number of points on

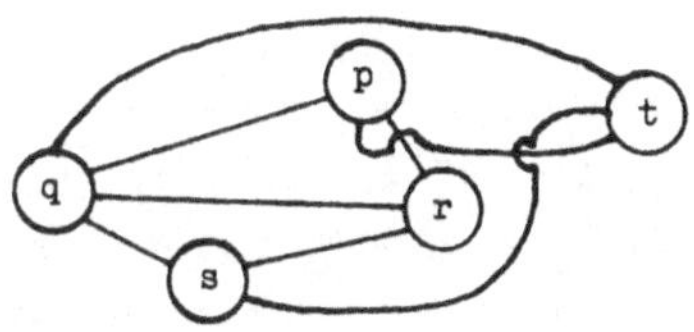

$w(p,r) = (p,r,s,t,q,p,r,s,\ldots)$
$m(p,r) = \langle p,r,s,t,q \rangle$
$w(q,r) = (q,r,p,t,s,q,t,p,q,r,\ldots)$
$m(q,r) = \langle q,r,p,t,s,q,t,p \rangle$

Fig. 1. Example of an oriented neighborhood structure.

a mesh m, i. e. the number of edges on this mesh, is called the *length of mesh* m, and denoted by $\lambda(m)$. For example, $\lambda(m(p, r)) = 5$ in Figure 1. In case of an isolated point $p \in P$, i. e. $\nu(p) = 0$, we define that $\langle p \rangle$ is considered to be a mesh of length zero, i. e. $\lambda(\langle p \rangle) = 0$.

Because of the unique successor relation for (directed) edges of a structure, any (directed) edge is contained in exactly one mesh. Let M be the set of all meshes of a structure (P, N, C). Then, the *Mesh Theorem*

$$\sum_{m \in M} \lambda(m) = 2\varkappa \qquad (2)$$

follows, and $\bar{\lambda} = 2\varkappa/\mu$ denotes the *averaged mesh length* of structure $\Gamma = (P, N, C)$, for $\mu = \mathrm{card}\,(M)$. In case of Figure 1, it holds that $\bar{\lambda} = 2 \cdot 8/3$.

Based on the definitions of neighborhood cycle and mesh, the following basic notions of image processing may be introduced: Let $\Gamma = (P, N, C)$ be a structure, and $Q \subseteq P$. A s u b s t r u c t u r e $\Gamma_Q = (Q, N_Q, C_Q)$ is defined by $N_Q = N \cap (Q \times Q)$, and C_Q is the restruction of cycles in C to points in Q.

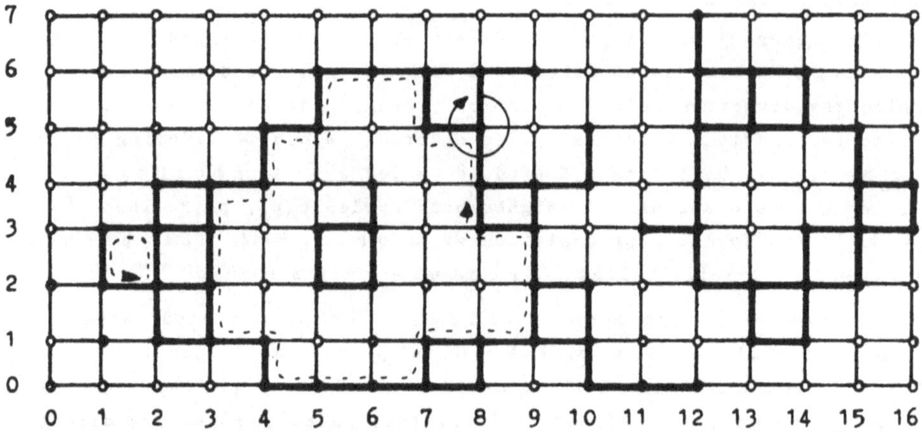

Fig. 2. A structure Γ (points o, •, edges —, ▬) and a structure Γ_Q (points •, edges ▬), where points p_{ij} are identified by row number i and column number j.

For example, in Fig. 2 cycle $c(p_{58}) = \langle p_{68}, p_{59}, p_{48}, p_{57} \rangle$ is restricted to cycle $c_Q(p_{58}) = \langle p_{68}, p_{48}, p_{57} \rangle$. Let $m_Q = m_Q(p, q)$ be a mesh of substructure Γ_Q, for $p, q \in Q$ and $(p, q) \in N_Q$. Then, m_Q is called

(i) a c o r e m e s h of Γ_Q iff $m_Q = m(p, q)$, i. e. m_Q is a mesh of structure Γ, too (for example, $m_Q = \langle p_{22}, p_{32}, p_{31}, p_{21} \rangle$ in Figure 2),
(ii) or a b o r d e r m e s h of Γ_Q iff $m_Q \neq m(p, q)$, i. e. the edge (p, q) generates a mesh in Γ different from m_Q (for example, $m_Q = \langle p_{48}, p_{58}, p_{57}, \ldots, p_{38} \rangle$ in Figure 2).

For $q \in Q$ and $p \in P - Q$, if $q \in N(p)$ then $\{p, q\}$ is an (undirected) b o r d e r e d g e of Q, (p,q) is an (ordered) b o r d e r p a i r of Q, and both $\{p, q\}$ and (p, q) are a t t a c h e d t o the border mesh $m_Q(s, q) = m_Q(q, r)$, for cycle $c(q) = \langle s, s_1, \ldots, s_k, p, p_1, \ldots, p_l, r, \ldots \rangle$ of point q in structure Γ and cycle $c_Q(q) = \langle s, r, \ldots \rangle$ of point q in substructure Γ_Q. For example, the border edges $\{p_{56}, p_{57}\}, \{p_{56}, p_{66}\}, \{p_{55}, p_{56}\}, \{p_{45}, p_{55}\}, \ldots$ are attached to the border mesh $m_Q(p_{58}, p_{57}) = m_Q(p_{57}, p_{67}) = m_Q(p_{67}, p_{66}) = \ldots$ in Figure 2.

4. Euler's Theorem

Structures may be constructed by successively adding a single point or a single edge (connecting two points already existing within the structure) to a given substructure. In this way it may be proved by induction that the E u l e r T h e o r e m

$$\varepsilon - \kappa + \mu = \varkappa \leq 2 \tag{3}$$

holds for connected (!) structures. Also it follows that the E u l e r c h a r a c t e r i s t i c $\varkappa$ of a given connected structure may be equal to an even number +2, 0, -2, -4, ... only. In general, if the given structure $\Gamma = (P, N, C)$ consists of γ components then $\varkappa = \varkappa(\Gamma) \leq 2 \cdot \gamma$. In case of step-wise building up a structure we have $\Delta\varkappa \leq 0$, and in case of step-wise restriction (deletion of points and/or edges) we have $\Delta\varkappa \geq 0$.

A structure $\Gamma = (P, N, C)$ consisting of γ components is defined to be p l a n a r iff $\varkappa = \varkappa(\Gamma) = 2\gamma$. It follows that a finite graph is planar iff there exists an orientation C on this graph such that the resulting structure is planar. For example, the complete graph K_5 with $\varepsilon = 5$ and $\kappa = 10$ cannot be planar because $\varkappa = 2$ would lead to $\mu = 7$, i. e. $\bar{\lambda} = 2\kappa/\mu = 20/7$, what is impossible because of $\lambda(m) \geq 3$ for any mesh m of a structure defined on K_5.

A structure $\Gamma = (P, N, C)$ is defined to be t o r o i d a l iff Γ is connected and $\varkappa = \varkappa(\Gamma) = 0$. For example, the structure of Figure 1 is toroidal.

A structure $\Gamma = (P, N, C)$ is h o m o g e n e o u s iff it is connected, for any point $p \in P$ we have $\nu(p) = \bar{\nu}$, and $\lambda(m) = \bar{\lambda}$ for any mesh m of the structure, i. e. in a homogeneous structure all points, edges and meshes are "enjoying the same rights". A homogeneous structure is called a n e t . The structure in Figure 1 in not homogeneous as well as the structure in figure 2 (because of "border points" with $\nu(p) < 4$). In the case of nets, the three basic formulas (1), (2), (3) of digital topology may be stated in the following form:

$$\varepsilon = 2\kappa/\nu , \qquad (4.1)$$

$$\mu = 2\kappa/\lambda , \text{ and} \qquad (4.2)$$

$$2/\nu + 2/\lambda = 1 + \varkappa/\kappa . \qquad (4.3)$$

Nets are the typical s u p p o r t s t r u c t u r e s of images used in digital image processing, i. e. grey value distributions are mapped onto nets for generating the typical subject of a "digital image of a certain size m x n". Therefore, nets and substructures of nets are worth some special attention. We shall briefly demonstrate how these three basic equations of nets, and notions related to these three equa-tions may be used to deduce theoretical results of practical importance, basic algorithms, and measures and features of digital point sets which may be considered to be "object areas".

5. Theoretical Results

The diophantic equa-tional system (4) contains six variables. By respecting the secondary conditions $\varepsilon, \kappa, \mu \geq 2$, $2 \leq \nu \leq \varepsilon - 1$, $\lambda \geq 3$, and $\kappa \leq \varepsilon(\varepsilon - 1)/2$, for a given Euler characteristic $\varkappa$ all solution tuples $(\nu, \lambda, \varepsilon, \kappa, \mu)$ of integers may be calculated. In the case of planar nets, i. e. for $\varkappa = 2$, only the five platonic tuples (3, 3, 4, 6, 4), (3, 4, 8, 12, 6), (3, 5, 20, 30, 12), (4, 3, 6, 12, 8), and (5, 3, 12, 30, 20) are integer solutions to (4). Also in case of $\varkappa < 0$, for any fixed value of $\varkappa$ there exist a finite number of nets for all of them the number of points ε is small relatively to the size of images normally processed in digital image processing. But note that in the case $\varkappa = 2$ and $\kappa \to +\infty$ the equation

$$2/\nu + 2/\lambda = 1 \qquad (4.3)^*$$

has three integer tuples as solutions: $(\nu, \lambda) = (4,4)$ for the infinite rectangular net, $(\nu, \lambda) = (3,6)$ for the infinite hexagonal net, and $(\nu, \lambda) = (6,3)$ for the infinite triagonal net.

Most interesting are toroidal nets, i. e. it is $\varkappa = 0$. In this case the three solutions of $(4.3)^*$ are characterizing finite rectangular, finite hexagonal, and finite triagonal nets, respectively. For $(\nu, \lambda) = (4,4)$, $(3,6)$, or $(6,3)$ there exist infinitely many nets with $(\nu, \lambda, \varepsilon, \varkappa, \mu) = (\nu, \lambda, 2\varkappa/\nu, \varkappa, 2\varkappa/\lambda)$ where the point number $\varepsilon = 2\varkappa/\nu$ exceeds any finite bound. On the other hand, these restrictions to $\nu, \lambda, \varkappa$ are sufficient only for proving the existence of a net with parameter set $\nu, \lambda, \varepsilon, \varkappa, \mu$. For example, conditions as $\varepsilon \geq \max\{\nu + 1, \lambda\}$, or $2\varkappa/\nu$ and $2\varkappa/\lambda$ have to be integers also must be satisfied. It may be proved that $(\nu, \lambda, \varkappa) = (4, 4, 10)$ is a toroidal net with the minimal value $\varepsilon = 5$, i. e. there exists an orientation on K_5 such that a net is given. It holds that for any integers $\alpha, \beta \geq 3$ there exists a rectangular net containing $\alpha \cdot \beta$ points exactely.

By these results it may be stated that in case of digital images of size m x n there exist rectangular, hexagonal, or triagonal toroidal nets such that planar substructures of these nets may be used as support structure for the given digital images of size m x n, where the point number ε of the torodial net has to be chosen sufficiently large. That means, there is no direct obligation from that point of view of embedding digital images into neighborhood structures, that *infinite* structures have to be assumed.

The following *Border Mesh Theorem* may be proved for border meshes of proper substructures $\Gamma_Q \subset \Gamma$: Let Γ^* be constructed out of Γ by deletion of all border edges which are attached to the same border mesh of Γ_Q. Then $\varkappa(\Gamma^*) > \varkappa(\Gamma)$. Because this Theorem is true for planar structures especially the *Separation Theorem* follows: Let Γ^* be constructed out of a planar connected structure Γ by deletion of all border edges which are attached to the same border mesh of a proper substructure Γ_Q. Then Γ^* consists of two (non-connected) components at least.

This Seperation Theorem represents the discrete analog of the Jordan Curve Theorem in the real plane. Any border mesh of a region $R \subset P$ of a planar connected structure $\Gamma = (P, N, C)$ separates region R and at least one of its complementary components of P - R. Thus, the Separation Theorem is the theoretical basis of any image segmentation. Because this Separation Theorem is not true in case of non-planar structures, as for example for the so-called " 8-neighborhood", the use of non-planar support structures should be avoided in image processing. Especially for the triangular net, i. e. the so-called " 6-neighborhood", it may be stated that any border mesh of a region of a planar substructure of a triangular net separates exactly one complementary component. Thus, any subset Q of a planar substructure of a triangular net generates a *region adjacency structure* (see Section 9) which is isomorphic to a tree. This is not true in general for rectangular or hexagonal nets where non-tree structures may result.

6. Algorithm s

The definitions of connectedness of subsets and of regions are directly leading to a solution to one of the most important problems in dealing with digital images: For a point set $Q \subseteq P$ all or some components have to be determined. Algorithmic solutions to this problem appear in several variants (row-by-row processing, neighborhood extensions, border line following etc.), and are basic modules of methods for image segmentation [16]. The basic idea of some of these solutions may be described as follows: For $Q \subseteq P$ a certain component has to be extracted starting at a certain point $q_0 \in Q$. Let R be initially equal to $\{q_0\}$. Then further points $q \in Q$ will be added to the connected set R if there is a point $r \in R$ with $q \in N(r)$, as long as such points q may be found. Despite of the special algorithmic solution with regard to a selected list structure or basic method (e. g. recursion) it may be noted that this approach may be realized without using the orientation C of the given structure $\Gamma = (P, N, C)$ but will be time-consuming because of its exhaustive nature.

An other class of solutions may be characterized by "border line following". Based on the definition of border meshes algorithms may be designed for calculating all the points or edges on a border mesh starting at a border edge attached to this border mesh. Then, using the neighborhood cycles of the points on the border mesh the border mesh may be constructed successively, and all the border edges attached to this border mesh may be determined simultaneously.

In this paper it should be stated that such algorithms may be designed at a very general level of abstraction. To illustrate this let us consider the following algorithm for computing border meshes. For a structure $\Gamma = (P, N, C)$, let (p, r) be a border pair of region $R \subset P$, i. e. $r \in R$ and $p \in P - R$ and $p \in N(r)$. There is a unique border mesh m of R such that (p, r) is attached to m. To calculate m proceed as follows:

(i) The ordered pair (p, r) is attached to mesh m. Initialize $(p, r) \rightarrow (q_0, r_0)$, $0 \rightarrow i$, and $0 \rightarrow j$.

(ii) The neighborhood cycle of point r_i is $c(r_i) = \langle \ldots, q_j, q, \ldots \rangle$. If $q \notin R$ then go to step (iv).

(iii) Point q is a further point of border mesh m. Let $i + 1 \rightarrow i$, and $q \rightarrow r_i$. The neighborhood cycle of point r_i is $c(r_i) = \langle \ldots, r_{i-1}, q, \ldots \rangle$. If $q \in R$ then go to step (iii).

(iv) If $(q, r_i) = (q_0, r_0)$ then STOP, else let $j+1 \rightarrow j$, $q \rightarrow q_j$, and go to step (ii).

If we leave the algorithm at STOP then the starting pair (p, r) was reached again. Note that coming back to point r is not sufficient (compare point q in m(q, r) in figure 1). The border mesh m is equal to $\langle r_0, r_1, \ldots, r_i \rangle$.

This algorithm may be used in any structure $\Gamma = (P, N, C)$ for computing border meshes of regions $R \subset P$. In the special case of substructures of rectangular nets (or more specific: of orthogonal rectangular grids) this algorithm reduces to the well-known border line following method [16], the calculated border lines are separating the "interior" of the given region from a component of its "exterior". But, this algorithm also may be used for very different structures as non-planar or non-homogeneous structures, too (compare Section 9 for examples of non-homogeneous structures). This remark points to a general advantage of the theoretical approach presented in this paper: A special structure that has to be considered within a special field of application of image processing may be characterized by the notions of the presented theory (planar or not, toroidal or not, homogeneous or not, connected or not etc.). Then, the proper subset of theorems, rules, formulas, algorithms etc. may be applied to the given structure. For example, a window of the orthogonal rectangular grid is a connected, planar substructure of a homogeneous oriented neighborhood structure. Thus, the algorithm for computing border meshes given above may be applied in this special situation for separating regions from its complementary components. The correctedness of this separation follows by the theoretical considerations before (Separation Theorem).

7. Curvature of Border Meshes

Let $\Gamma = (P, N, C)$ be a toroidal net with $(\nu, \lambda) = (4, 4)$, $(3, 6)$, or $(6, 3)$. Assume R to be a region with a single border mesh of length 1 only, i. e. R has no "holes". The number of border edges attached to this border mesh be n. Then, the three topological basic equations are modified to

$$\nu\varepsilon - n = 2\varkappa, \tag{5.1}$$

$$\lambda(\mu - 1) + 1 = 2\varkappa, \text{ and} \tag{5.2}$$

$$\varepsilon - \varkappa + \mu = \varkappa(\Gamma_R), \tag{5.3}$$

for the substructure $\Gamma_R = (R, N_R, C_R)$. The first equation follows because each of the ε points of R has ν neighbors exactly, but n of these neighbors are not contained in the set R. The equation (5.2) holds because all of the $\mu - 1$ core meshes of Γ_R have length λ, and the single border mesh has length 1 per definition. Let $\varkappa(\Gamma_R) = \varkappa_R$. Of course, only the values $\varkappa_R = 0$ or $\varkappa_R = 2$ are possible. The expression

$$\nu\lambda \cdot \text{equation (5.3)} - \lambda \cdot \text{equation (5.1)} - \nu \cdot \text{equation (5.2)}$$

leads to

$$\lambda n - \nu 1 = \nu\lambda(\varkappa_R - 1) + (2\lambda + 2\nu - \nu\lambda)\varkappa .$$

In toroidal nets it holds that $2\lambda + 2\nu - \nu\lambda = 0$, thus this equation may be transformed to

$$g = \frac{n}{\nu} - \frac{1}{\lambda} = \varkappa_R - 1 \tag{6}$$

where g is called the t o t a l c u r v a t u r e of the single border mesh of Γ_R. The total curvature may be equal to +1 in case of planar regions, or equal to −1 in case if toroidal regions, the border mesh of the region is called an o u t e r b o r d e r m e s h , or an i n n e r b o r d e r m e s h respectively.

Now assume that region R has r border meshes $m_1, m_2, \ldots, m_r$. Let l_i be the length of mesh m_i, and let n_i be the number of border edges attached to mesh m_i. Then, similarty to (5) we obtain

$$\nu\varepsilon - \sum_{i=1}^{r} n_i = 2\varkappa , \tag{7.1}$$

$$\lambda(\mu - r) + \sum_{i=1}^{r} l_i = 2\varkappa , \text{ and} \tag{7.2}$$

$$\varepsilon - \varkappa + \mu = \varkappa_R \tag{7.3}$$

for the substructure $\Gamma_R = (R, N_R, C_R)$. Let $n^* = \sum_{i=1}^{r} n_i$ and $l^* = \sum_{i=1}^{r} l_i$. It followm that

$$r = \frac{n^*}{\nu} - \frac{l^*}{\lambda} + \varkappa_R , \tag{8}$$

i. e. the global topological structure characterized by the value of r may be calculated by the values n^* and l^* which may be obtained by local topological considerations. Furthermore, based on equational system (7) the following characterizations of mesh curvature may be proved:

Any planar region R has exactly one outer border mesh of total curvature $\frac{n}{\nu} - \frac{l}{\lambda} = +1$. All the other border meshes of R are inner ones with total curvature $\frac{n}{\nu} - \frac{l}{\lambda} = -1$.

Because l and n may be calculated during the computation of a border mesh, the total curvature signalizes the type of border mesh at the end of this computational process. Finally, equation (6) may be used to define the *interior* or *exterior* relatively to the computed border mesh.

8. Pick Formulas

Let R be a planar region of a toroidal net without inner border meshes. For $\frac{n}{\nu} - \frac{l}{\lambda} = +1$ and $f = \mu - 1$, the number of core meshes of R, the expression

equation (5.1) = equation (5.2)

leads to the equation

$$\varepsilon = \frac{\lambda}{\nu} f + \frac{l}{2} + 1. \tag{9}$$

This equation may be characterized to be the topological generalization of the formula $A = i + \frac{b}{2} - 1$ found by G. Pick [17], where A denotes the area content of a simple grid point polygon within the rectangular grid (each cell is assumed to have area content equals 1), i denotes the number of grid points in the interior of this polygon, and b denotes the number of grid points on the border of this polygon.

In the case of an inner border mesh a similar equation

$$\varepsilon = \frac{\lambda}{\nu} f - \frac{l}{2} + 1 \tag{10}$$

may be proved, where ε denotes the number of points enclosed by the border mesh and not on that mesh, f denotes the number of meshes of the net enclosed by the border mesh, and l denotes the length of the border mesh.

Formulas of this type may be used in digital image processing for calculating topological values. For example, let us assume that within a connected, planar substructure of a toroidal (ν, λ) - net $\Gamma = (P, N, C)$ there are γ components of a subset $Q \subset P$ exactly all without inner border meshes. This situation is typical in case of binary digital images, e. g.. Then, by adding all the equations $\frac{n_i}{\nu} - \frac{l_i}{\lambda} = +1$, for $i = 1, 2, \ldots, \gamma$, we obtain an equation

$$\frac{n^*}{\nu} - \frac{l^*}{\lambda} = \gamma . \tag{11}$$

The values $n^* = \sum_{i=1}^{\gamma} n_i$ and $l^* = \sum_{i=1}^{\gamma} l_i$ may be obtained by local considerations within the given digital image. Then equation (11) gives the number of components of Q without performing an image segmentation process. Similar characterizations of global by local situations were demonstrated in [2, 6, 12] for calculating the so-called Euler number. - An other way to use formulas (9) and (10) may be connected with computing the area content of regions.

9. Non-homogeneous Structures

Despite of the homogeneous support structures $\Gamma = (P, N, C)$ for defining digital images by mapping grey values onto the points of connected, planar substructures of such a structure, different types of non-homogeneous structures are relevant to digital image processing. For sets $Q \subset P$, substructures Γ_Q are non-homogeneous and were considered in the Sections before. Further examples are region adjacency graphs, and mosaic structures.

Finite Poisson grid-structures: Let us assume that n straight lines are (randomly) placed in the Euclidean plane such that there is no pair of parallel straight lines, and such that no three straight lines are passing through the same point. Finally, assume that all 2 n rays will be deleted such that a bounded figure results which is characterized by $\varepsilon = n(n-1)/2$ crossing points, $\varkappa = n(n-2)$ straight line segments (connecting crossing points), and $\mu = 2 + \frac{n(n-3)}{2}$ faces or cells (including the unbounded exterior). For the averaged valence of this structure, we obtain $\bar{\nu} = 2\varkappa/\varepsilon = 4 \cdot \frac{n-2}{n-1}$ (which goes to 4 for $n \to \infty$), and for the averaged mesh length it follows that $\bar{\lambda} = 2\varkappa/\mu = 4 \cdot \frac{n(n-2)}{n(n-3)+4}$ (which also goes to 4 for $n \to \infty$).

Finite Voronoi structures: Let us assume that n points $p_1, p_2, \ldots, p_n$ are (randomly) placed in the Euclidean plane such that no two points are identical, and such that no four points are cocircular, i. e. on the same circular line. Then $V(p_i) = \{q \mid d(q, p_i) \leq d(q, p_j), \text{ for } j = 1, 2, \ldots, n\}$ denotes the Voronoi cell of point p_i which in a convex polygonal area, where d is the Euclidean metric. All these bounded or unbounded cells $V(p_1), \ldots, V(p_n)$ are defining an (unbounded) Voronoi grid. Finally, by deletion of all the k rays of this grid starting at a certain grid point, a finite Voronoi structure results which is characterized by $\varepsilon = 2n - k - 2$ grid points, $\varkappa = 3n - 2k - 3$ straight line segments (connecting grid points), and $\mu = n - k + 1$ faces or cells (including the unbounded exterior).

Note that k is the number of external points of the convex hull of the set $\{p_1, p_2, ..., p_n\}$. For the averaged valence of this structure, we obtain $\bar{v} = 2\kappa/\varepsilon = (6n - 4k - 6)/(2n - k - 2) < 3$ (which goes to 3 for $n \to \infty$ because k/n goes to zero for $n \to \infty$), and for the averaged mesh length it follows that $\bar{\lambda} = 2\kappa/\mu = (6n - 4k - 6)/(n - k + 1)$ (which goes to 6 for $n \to \infty$).

Region adjacency graphs: For a given neighborhood structure (P, N) let $\underline{R}$ be a partition of P into non-empty, pairwise disjoint regions. For $R_1, R_2 \in \underline{R}$ we define $R_1 \in \underline{N}(R_2)$ iff there exist points $p_1 \in R_1$ and $p_2 \in R_2$ such that $p_1 \in N(p_2)$. Then, the neighborhood structure $(\underline{R}, \underline{N})$ is called a c o n c e n t r a t e of the neighborhood structure (P, N). This operation of concentration may be performed repeatedly, and the resulting neighborhood structures are isomorphic to the so-called region adjacency graphs. For any neighborhood structure (P, N) isomorphic to a planar graph there exists an orientation C such that (P, N, C) is a planar structure. Because the operation of concentration may be considered to be a sequence of restrictions of this structure (P, N, C), we have $\Delta\chi \geq 0$. Thus, the resulting structure $(\underline{R}, \underline{N}, \underline{C})$ is planar, i. e. the neighborhood structure $(\underline{R}, \underline{N})$ is planar in the sense of graph theory. As a special case, $\underline{R}$ may consist of all components of a subset $Q \subseteq P$, and of all complementary components of this set Q, and let $\underline{R}_Q = \underline{R}$ in this special case. As a special consequence of the three topological basic equations, if (P, N, C) is a toroidal (6, 3) - net then any concentrate $(\underline{R}_Q, \underline{N})$ is isomorphic to a tree what is not the case for toroidal (4, 4) - or (3, 6) - nets (see Figure 3). The process of concentration may be designed accordingly to graph theoretic properties. For example, the vertices labeled by R_{65} and $R_{7,12}$ are a r t i c u l a t i o n p o i n t s of the graph in Figure 3 because after the deletion of such a point the resulting graph will consist of more components than the graph before.

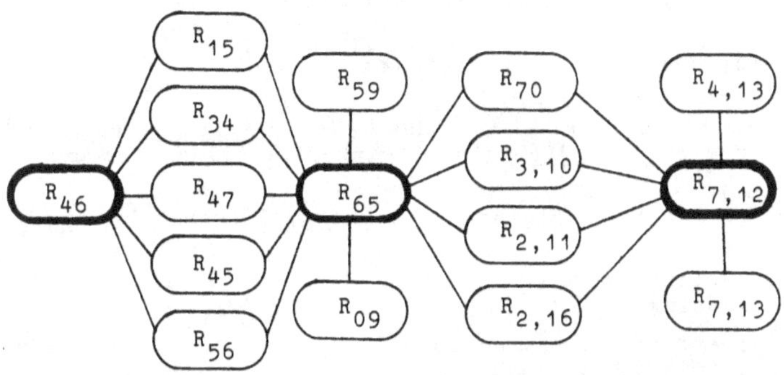

Fig. 3. Region adjacency graph of the neighborhood structure $(\underline{R}_Q, \underline{N})$ defined by the substructure Γ_Q shown in figure 2. Region R_{ij} contains point p_{ij} in its uppermost, leftmost corner. It is $Q = R_{46} \cup R_{65} \cup R_{7,12}$.

10. Conclusions

In this paper some examples of applications of the Vertex Theorem (1), the

Mesh Theorem (2), and the Euler Theorem (3) were given. To our opinion, these three equations may be called the *topological basic equations* of neighborhood structures because of their fundamental importance. The application of these equations in the context of special conditions, as demonstrated by the equations (4), (5), and (7), leads to a wide variety of results which are of direct importance to digital image processing.

The point of view of the Theoretical Physics may be applied to motivation: The Mechanics is based on the Newton equation, the Quantum Physics on the Schrödinger equation, the Electro Dynamics on the Maxwell equations. Only by tracing back to a few fundamentals a concentrated representation of the field may be obtained which is essentially different from a selection of "cooking recipes".

In comparison to the "high level vision problems" the theory dealt with in this paper is located at the very ground level. But a well-build base is always a good pledge. And, still there are important open problems at this ground level. For example, is there a three-dimensional analog to the two-dimensional oriented notion of a mesh? Is it possible to characterize "three-dimensional planarity or regularity" which is a property of the three-dimensional orthogonal grid, e. g., but not of the "8-neighborhood"?

References

1. H. Freeman, On the encoding of arbitrary geometric configurations. IRE Trans. EC - 10 (1961), 260-273.
2. M. Minsky, S. Papert, Perception - An Introduction to Computational Geometry. MIT Press, 1969.
3. A. Rosenfeld, Connectivity in digital pictures. JACM 17 (1970), 146 - 160.
4. J. Sklansky, Measuring concavity on a rectangular mosaic. IEEE Trans. on Comput. C - 21 (1972), 1355 - 1361.
5. H. Simon, K. D. Kunze, K. Voss, W. R. Herrmann (eds.), Automatische Bildverarbeitung in Medizin und Biologie. Steinkopff-Verlag, Dresden, 1975, pp. 66 - 100.
6. A. Rosenfeld, A. C. Kak, Digital Picture Processing, Academic Press, New York, 1982.
7. R. Klette, The m-dimensional grid point space. Computer Vision, Graphics, and Image Processing 30 (1985), 1 - 12.
8. V. A. Kovalevski, Discrete topology and contour definition. Pattern Recognition Letters 2 (1984), 281 - 288.
9. A. Hübler, Zur mathematischen Grundlegung der diskreten Geometrie. Kurzreferate 1. Intern. Fachtagung "Automatische Bildverarbeitung", Berlin (DDR), 17. und 18. Oktober 1985, A1/1 - A1/6.
10. C.-N. Lee, A. Rosenfeld, Continuous Representation of digital images. TR - 1569, Center for Automation Research, Univ. of Maryland, College Park, 1985.
11. G. Matheron, Random Sets and Integral Geometry, J. Wiley & Sons, New York, 1975.
12. J. Serra, Image Analysis and Mathematical Morphology, Academic Press, New York, 1982.
13. K. Voss, P. Hufnagl, R. Klette, Theoretische Grundlagen der digitalen Bildverarbeitung I. Bild und Ton 38 (1985), 299 - 302. See also subse-

quent papers: Bild und Ton 38 (1985), 325 - 331 and 39 (1986),45 - 50, 213 - 219, 303 - 307 and 40 (1987), Heft 4.

14. Enzyklopädie der Elementarmathematik, Band V, DVW, Berlin, 1971.

15. B. Klotzek, Kombinieren - Parkettieren - Färben. DVW, Berlin, 1985.

16. K. Voss, K. Roth, P. Hufnagl, Objektisolierung und Objektmerkmale. AUTBILD '84/1, Wiss. B eiträge FSU Jena, 1984, 98 - 138.

17. G. Pick, Geometrisches zur Zahlentheorie. in: Zeitschrift des Vereins "Lotos", Prag, 1899.

THE TOPOLOGY OF CELLULAR COMPLEXES AS APPLIED TO IMAGE PROCESSING

V. A. Kovalevsky[1)]

Abstract

The notion of a cellular complex which is well-known in the topology is applied to describe the structure of images. It is shown that such a description is in full accordance with the classical topology. No contradictions or paradoxes arise when defining connected subsets and their boundaries. Ways of encoding images as cellular complexes are discussed. The process of image segmentation is considered as splitting (in the topological sense) a cellular complex into blocks of cells. The notion of a cell list is introduced as a precise and compact data structure for encoding segmented images. Some applications of this data structure to the image analysis are demonstrated.

1. Introduction

An image may be regarded as a set of gray values. It is however obvious that the positions of the gray values relative to one another are of great importance. Means to define the corresponding spatial relations are subject of the topology. The modern general topology is mainly considering sets of space elements or points having the property that in an arbitrarily small neighborhood of a point there are infinitly many other such points. An example of such a set is the continuum. The number of its elemenst is uncountable.

A digitized image, to the contrary, has always a finite number of elements. Thus we need a consistent topology of finite sets. If one will try to look for this notion in a textbook on topology (e.g. /1/) one will find a structure of isolated points in which no neighborhood, no connectedness and no boundaries of subsets may be defined. Such a topology called the "discrete topology" is of no use for image processing.

On this reason the specialists for image processing were forced to look for their own solution of the problem. At the beginning of the seventies Rosenfeld /2/ has suggested to regard an image as a graph whose nodes are the pixels and whose branches are linking the neighboring pixels to each other. It is called the <u>neighborhood graph</u>. This notion is widely used in the image processing and computer graphics. Thus the pos-

---

1) Central Institute of Cybernetics and Information Processing, Kurstrasse 33, 1086 Berlin, GDR

sibility has arisen to define certain topological notions, e.g. that of connectivity, for finite sets.

However the attempts to develop a consistent topology of two-dimensional images by means of graphs have failed due to the well-known connectivity paradox and some other contradictions /3,4/. Especially unpleasant were the contradictions in the definitions of the boundaries:

- There existed definitions for the inside and the outside boundary, as well as for a 4-connected and 8-connected ones. Neither of them have led to a consistent definition of adjacent subsets. Moreover, a boundary such defined was itself a set of pixels and thus have had a finite area, while according to our intuition it must be a one-dimensional set having a zero area. Further, the boundary of a set and that of its complement were different. If a boundary have had a self-contact the direction of its tracing at such a point was ambiguous. It was impossible to transfere the most important topological notions such as that of an open subset, open neighborhood, continuity and many other to a two-dimensional structure described by a neighborhood graph.

Substantial steps to overcome the difficulties have been made in the seventiens: Rosenfeld /5/ has suggested using different neighborhoods for the background and the objects. He also noticed that the 6-neighborhood does not lead to the connectivity paradox. Pavlidis and Feng /3,6/ have introduced the notion of the "extended boundary" which has solved the region adjacency problem, Serra /7/ has suggested to consider a mixed 4/8-neighborhood, Elliott and Srinivasan /8/ have proposed to consider the so-called boundary elements i.e. line segments separating two adjacent pixels from each other, Klette and Voss /9/ have introduced the notion of "meshs" corresponding to cycles in the neighborhood graph which uniquely define its imbedding into the plane. All these improvements turn out to lead to a general conception considering a digital plane as a structure consisting of <u>heterogeneuos elements</u> namely of elements of different dimensions: 0-dimensional points, 1-dimensional line elements and 2-dimensional area elements /10,11/. Such a structure is well-known in the topology as a cellular complex.

2. The Topology of Cellular Complexes

This classic branch of the general topolgy was foundet by the famos Riemann in the 19th century. It considers structures consisting of elements of different dimensions. Let E be a set of abstract elements called the <u>cells</u>.

<u>Definition 1:</u> An abstract <u>cellular complex</u> C=(E,B,dim) /1/ is such a set E provided with an antisymmetric and transitive binary relation $B \subset E \times E$ called the <u>bounding relation</u> and with a <u>dimension</u> function dim: $E \rightarrow I$ from E into the set I of the non-negative integers.

The bounding relation B is a partial order in E. It indicates the ordered pairs (e',e") of elements such that e' is said to <u>bound</u> e", or e' is a <u>face</u> of e" which is denoted as $e' < e''$. This means that a cell may be a face of another cell of higher dimension in the same sense in which a triangle may be a 2-dimensional face of a tetrahedron, a line segment may be a 1-dimensional face of a triangle, or of a tetrahedron, and a point may be a 0-dimensional face of a line segment, or of a triangle, or of a tetrahedron.

The function dim defines the dimension of each element. It is monotone, i.e. from $e' < e''$ follows $\dim(e') < \dim(e'')$. An element e' with $\dim(e')=d$ is called a d-dimensional element or a <u>d-cell</u>. A complex is called k-dimensional or a <u>k-complex</u> if the dimension of all its elements is less or equal to k. Examples of a 2-complex and a 3-complex are shown in Fig. 1.

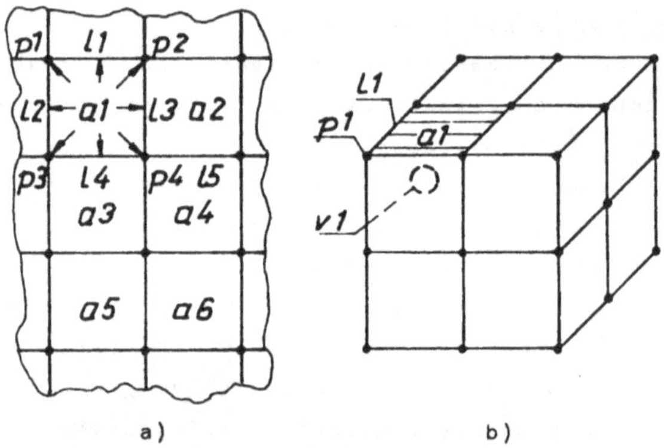

Fig. 1. Examples of a 2-dimensional (a) and a 3-dimensional (b) complexes

The 2-dimensional elements are the area elements. They must be associated with the notion of pixels since a gray value assigned to a pixel originates from measuring the amount of energy radiated from an elementary area.

A subcomplex C'=(E',B',dim') of a given complex C=(E,B,dim) is a complex whose set E' is a subset of E and the relation B' is an intersection of B with E'xE'. This means that the bounding relation remains the same as that of C for every pair of elements contained in E'. The dimensions also do not change: dim'(e)=dim(e) for all $e \in E'$. All this is important because it means that to define a subcomplex C' of C=(E,B,dim) it suffices to define the corresponding subset E' of the elements. Thus it is possible to speak of a subcomplex $E' \subset E$ while understanding the subcomplex C'=(E',B,dim). All subcomplexes of C may be regarded as subsets of E and thus it is possible to use the usual formulae of the set theory to define intersections, unions and complements of subcomplexes.

The defined hierarchy of elements of different dimensions makes it possible to consistently define the notion of an open finite subset:

<u>Definition 2:</u> A subcomplex S of C is called <u>open</u> in C if S contains all elements of C which are bounded by some element of S.

Thus S1={a1, l3, a2} is open in the complex of Fig. 1a but S2={l4, p4, l5} is not open in it since S2 does not contain e.g. the element a1 which is bounded by l4. Every open subset of C containing a given element e' of C is called an open neighborhood of e' in C. The open subset St(e',C) consisting of e' and all elements of C which are bounded by e' is called the open star of e' in C. It is the smallest open neighborhood of e' in C. Open stars of various elements in complexes of different dimensions are shown in Fig. 2.

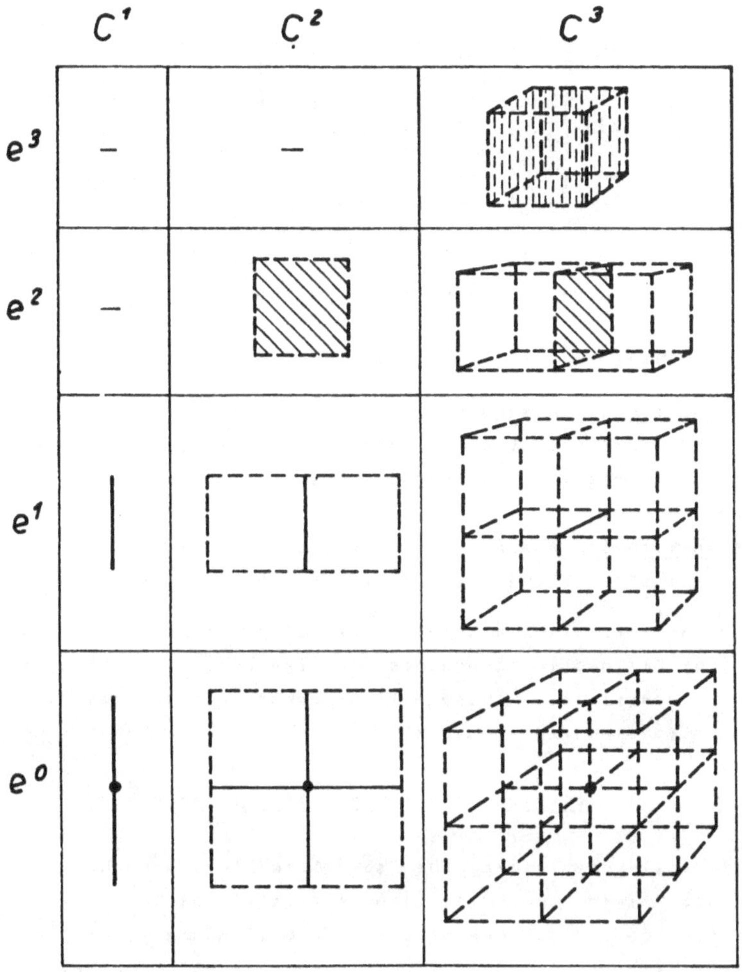

Fig. 2. The open stars of various elements in complexes of dimensions 1, 2 and 3

It is easy to see that the open subcomplexes (subsets) defined in this way satisfy the axioms of the topology. Therefore any complex is a topological space. All notions and results of the general topology may be transferred to complexes. A very important example gives the following

Definition 3: The boundary of a subset S in C is the subset Fr(S,C) of all elements e' of C such that any open neighborhood of e' contains elements both of S and of its complement C\S.

It is more convenient for the practice to replace in Definition 3 "any open neighborhood" by "an open star St(e',C)". Thus the subset S of Fig. 3a consisting of the elements a1, a2, a3, l1, l2 and p has a boundary shown in Fig. 3b by heavy lines and dots. It is easy to see that the open stars of all boundary elements satisfy Definition 3.

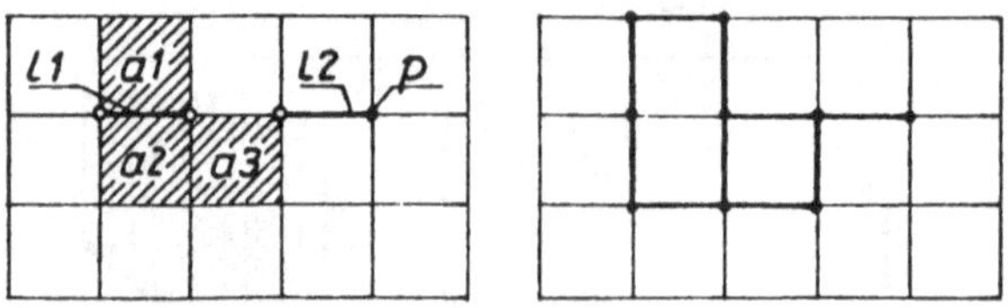

a) b)

Fig. 3. Example of a subcomplex (a) and its boundary (b)

Now we shall define the important notion of connectivity.

Definition 4: A sequence of elements of a subset S beginning with e' and finishing at e" is called a path in S from e' to e" if for every two elements which are adjacent in the sequence one of them is bounding the other.

Definition 5: A subset S is called connected if for any two elements e' and e" of S there exists a path in S from e' to e".

As one may see, all the topological notions important for image processing may be defined in accordance with the general topology and therefore they are consistent. These notions constitute the way of abolishing the contradictions mentioned above.

3. Solution of the Connectivity Paradox

It is well-known that the so-called 4-neighborhood and the 8-neighborhood lead to the connectivity paradox /3,4/. The 6-neighborhood however is free of the paradox. Hence it may seem that in a consistent finite toplogical space every 2-cell must have 6 adjacent 2-cells. But the classical topology is considering complexes whose 2-cells may have an arbitrary number of "neighbors" without running against contradictions. The explanation of this phenomenon consists in the following /11/. The connectivity paradox disappears if every subset of the complex representing an image is correctly defined. This means one should never define a subset S by simply specifying all pixels belonging to it and saying nothing about the space elements of lower dimensions. A correct definition may be done by a membership predicate M(S,e) according to which any space element of every dimension is declared as belonging either to S or to its complement C\S. Especially plausible is the case when N subsets Si, i=1,2,...N compose a partition of the space C. Then any membership predicate may be interpreted as coloring every space element of any dimension by one of the N colors.

Consider a rather illustrative example of Fig. 4.

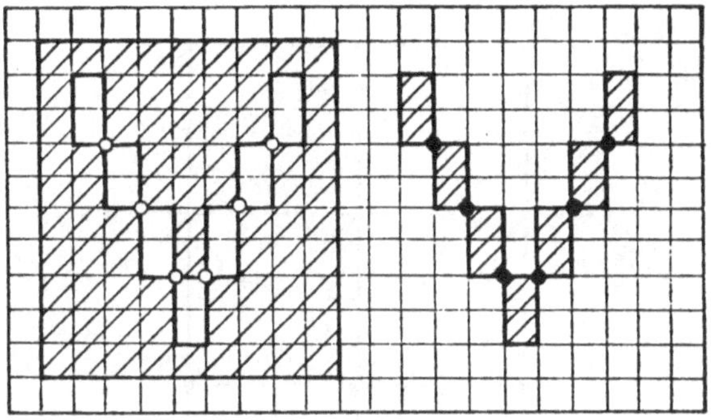

Fig. 4. Example of a consistent coding of connected black and white diagonals

Neither of the commonly used neighborhood relations makes it possible to define the boundaries of both V-shaped regions of Fig. 4 in such a way that the boundary of each V be connected and a tracing procedure performs in both cases similar sequences of steps. It is however easily realizable if we define the 0-cells depicted by white circles as being members of the white V-shaped region and the 0-cells depicted by black circles as belonging to the dark V. Then each boundary must be traced in such a way that both the 0-cells and the 2-cells having the color of the chosen region remain always to the right hand side of the direction of the tracing. No connectivity paradox arises in the case of the white V since the white pixels are connected to each other through the white 0-cells but the black pixels of the background are disconnected. A similar situation takes place for the black V.

Thus we see that the paradoxes vanish ones for all if we cease forcing a neighborhood relation to pairs of space elements of equivalent dimension. The only consistent way of specifying the connectivity of some subsets of an image is that of representing the image as a cellular complex and defining the membership of all cells of any dimension in these subsets. Such a representation of images not only abolishs the contradictions. It also gives the possibility to adequately represent the structure of images containing fine details as it was demonstrated by the example of Fig. 4.

The coding of the membership of the elements of lower dimensions must be realized in the same way as that of the high dimensional ones (the pixels): a memory word or a part of a word must be assigned to each element. If a part of a memory word is sufficient to encode the membership then the 0- and 1-dimensional elements may be assigned each to a certain 2-dimensional element. The memory word corresponding to a pixel must then be shared among the pixel and the elements assigned to it. An example of such assignment is shown in Fig. 5.

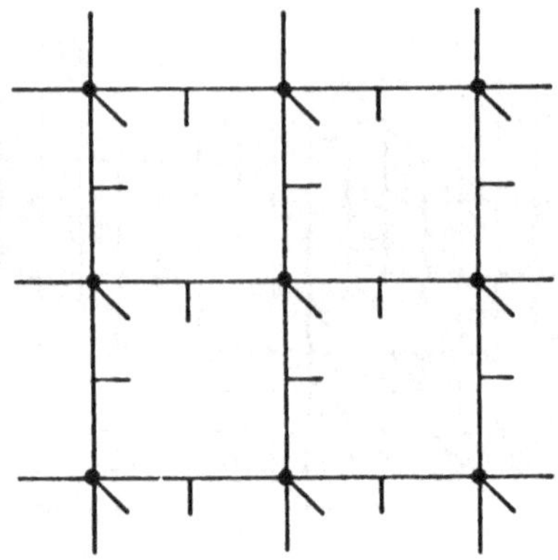

Fig. 5. One possible assignment of the 0- and 1-dimensional elements to the 2-dimensional ones

Similar assignments may be declared in the 3-dimensional case.

It is evident that there are many ways to use a priori knowledge about the sets under consideration to obtain economical coding of the set membership. It is quite natural e. g. to try introducing some kind of a <u>standard face membership</u>, i.e. a rule specifying the set membership of the faces of every 2-cell as a function of the membership of the 2-cell itself. However such a standard face membership cannot be chosen arbitrarily. It must be representable as a membership predicate as defined above. It is not allowed e. g. to declare all faces of every 2-cell as belonging to the same set S as the 2-cell itself since one and the same element may be a face of two 2-cells one belonging to S and the other to C\S. An example of a consistent standard face membership may be given by declaring every 0- and 1-cell as belonging to the same set as the 2-cell uniquely defined by the assignement rule of Fig. 5. It is easy to see that such a standard face membership is equivalent to declaring a 6-neighborhood of the 2-cells. This may serve as the explanation why the 6-neighborhood is free of the connectivity paradox.

The two other most often used neighborhoods may also be regarded as special cases of a standard face membership: the 4-neighborhood may be obtained if one regards the 0-cells as <u>non-existent</u>, i.e. every 0-cell belongs neither to S nor to its complement C\S, which is impossible. The 8-neighborhood corresponds to <u>doubled</u> 0-cells: every 0-cell exits in two copies one being in S and the other in C\S, which is also impossible. Hence both the 4- and the 8-neighborhood lead to a contradiction and should not be used any more. There is absolutely no need of using them since any wanted connectedness relation may be consistently defined by a corresponding membership of all cells of the space as it was demonstrated by the example of Fig. 4.

4. Splitting a Complex into Blocks

An image may be regarded as a mapping of a two-dimensional complex (the raster) onto a one-dimensional complex (the gray scale) or onto a set of symbols. Thus to define an image a gray value must be assigned to every raster element. This assignment is most often performed during the digitization of optical images: a fragment of the euclidean plane is being dissected into small area pieces, the brightness of each piece is being measured and the result assigned to the corresponding pixel of the raster. The membership of the 0- and 1-cells may be defined by some standard face membership rule as explaned above. The rule must use the available a priori knowledge about the class of images under consideration.

The well-known procedure of image segmentation may be regarded from the point of view of cellular complexes as <u>splitting</u>: a cellular complexes obtained as the result of the digitization is being splitted into blocks of cells having all approximately one and the same gray value or color. These are the <u>regions</u> satisfying the "uniformity predicate" introduced by Pavlidis /3/. The term "region" is in accordance with the corresponding notion of the topology where a region is defined as an open connected two-dimensional subcomplex.

The boundary of a region is a one-dimensional subcomplex which may have <u>branching points</u> at the locations where three or more regions meet. Parts of a boundary separated by branching points are called <u>boundary segments</u>. There exits a bounding relation between these objects: the boundary segments are bounded by some branching points, the regions are bounded by some boundary segments and branching points. Thus the collection of these objects may be regarded as a new cellular complex whose 2-cells are the regions, the 1-cells are the boundary segments and the 0-cells are the branching points. These new cells are the blocks obtained as the result of splitting the original complex. We shall call such a new complex the <u>block complex</u>. An example of a block complex is shown in Fig. 6.

A block complex BC may be regarded as a generalization of the notion of a <u>region adjacency graph</u> /5,3/. The latter is from the topological point of view the 1-skeleton of the complex dual to BC. The block complex exactly represents the topology of the original image. This is its important advantage as compared to the region adjacency graph.

A block complex may be described by a data structure called the <u>cell list</u>. It consists of three lists: that of its 0-dimensional, 1-dimensional and 2-dimensional cells. To provide the possibility of

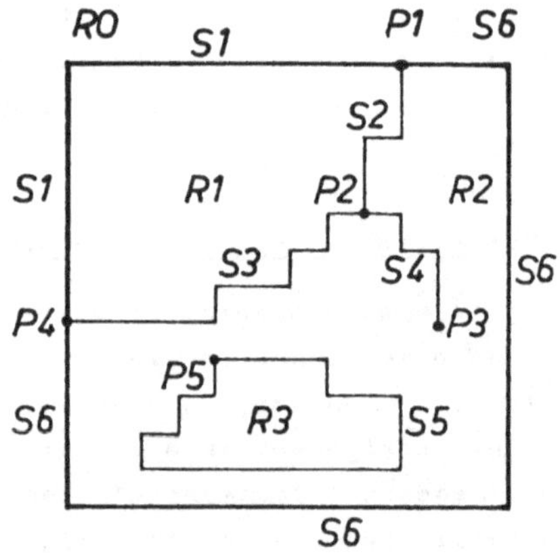

Fig. 6. Example of a block complex. Ri are the 2-cells, Si are the 1-cells and Pi are the 0-cells

analyzing and restoring the original image the information represented in the lists of cells must be augmented by the <u>metric</u> <u>data</u>: the coordinates of the 0-cells and some intermediate points in the 1-cells which are gained during the boundary tracing.

The tracing is accompanied by the recognition of <u>digital</u> <u>straight</u> <u>segments</u> (DSS). It provides an efficient polygonal representation of the boundaries. We have succeeded in encoding the DSS in such a way that an exact reconstruction of the original boundaries be possible. Thus our polygonal representation is an exact and economical encoding of the boundaries rather than an approximation. We are using an original fast method of recognizing the DSS /12/.

The structure of the cell list may be explained by the example of the image of Fig. 6. The corresponding lists are shown in Table 1. In the list of the 0-cells the first column contains the identifiers of the 0-cells used in Fig. 6. The column "val" indicates the number of the 1-cells bounded by the corresponding 0-cell. The next 6 columns contain the identifiers (pointers) of all cells bounded by the current 0-cell.

In the list of the 1-cells the last 2 columns contain pointers to the subsequence of the coordinate pairs corresponding to the intermediate points of the 1-cell. In the same way the contents of the last 2 columns of the list of 2-cells are pointers to the list of boundary sequences. Each sequence contains all 1-cells composing the boundary of the 2-cell. The minus sign indicates that the 1-cell is oriented against the circulation of the boundary. The labels in the second column of the list of the 2-cells indicate the gray value, or the color, or some other features of the 2-cells.

Table 1

The list structure for the cellular complex of Fig. 6

List of the 0-cells

No	X	Y	val	1-cells			2-cells		
P1	9	0	3	S6	S2	S1	R2	R1	R0
P2	8	4	3	S4	S3	S2	R2	R1	R2
P3	10	7	1	S4	-	-	R2	-	-
P4	0	7	3	S3	S6	S1	R2	R0	R1
P5	4	8	2	S5	S5	-	R3	R2	-

List of the 1-cells

No	0-cells		2-cells		metric	
	begin	end	left	right	beg.	end
S1	P1	P4	R1	R0	1	1
S2	P1	P2	R2	R1	-	-
S3	P2	P4	R2	R1	2	2
S4	P2	P3	R2	R2	-	-
S5	P5	P5	R2	R3	3	5
S6	P4	P1	R2	R0	6	8

List of the 2-cells

No	label	bound. seq.	
		begin	end
R0	L0	1	2
R1	L1	3	5
R2	L2	6	10
R3	L3	11	11
-	-	-	-
-	-	-	-

List of the metric data

address:	1	2	3	4	5	6	7	8
coord.:	(0,0)	(4,7)	(9,9)	(9,11)	(2,11)	(0,12)	(12,12)	(12,0)

List of the boundary sequences

address:	1	2	3	4	5	6	7	8	9	10	11
pointer:	S6,	S1,	-S1,	S2,	S3,	-S2,	-S6,	-S3,	S4,	-S4,	S5

5. Applications

The described technique was implemented on the image processing systems ROBOTRON A6471 and A6472 /13/. The latter contains a fast pipe-line processor which is able to label all 0-cells of a previously segmented image of 256x256x8 bits by the "direction labels" in a time of less than 1 sec. The labels indicate the direction of the next step when tracing the boundary. Therefore the boundary tracing (which cannot be realized by the "pseudo-parallel" processor of the A6472) becomes also fast.

The program scans the labeled image line by line, looks for new boundaries and starts the tracing procedure which is combined with the recognition of the digital straight segments. Both routines make records in the cell list whose structure was described in the previous section.

The cell list for an average image of 256x256x8 bits containing up to 120 regions having the form of narrow stripes, rectangles, rings, crosses, is produced by a rather slow mini-computer in about 70 sec.

The metric data contained in the list make any geometrical calculations especially efficient and simple: the distance between two points, the area of a region, the perimeter of a boundary etc. may be calculated by the usual well-known means dealing with the coordinates. All geometrical transformations e.g. translation, scaling, rotation etc. may be performed by recalculating all the cooorditates of the 0-cells and the intermediate points according to the well-known formulae. To obtain the transformed image in the raster representation the list structure must be converted back into the raster format. Experimental results demonstrating the advantages of this way of geometrical transformations are presented by N. Luth /14/.

The further processing of the cell list consists in recognizing simple geometrical figures such as the ones mentioned above. The possibility to analyse the coordinates of polygon vertices instead of carrying out a mask matching makes the recognition extremely tolerant to geometrical transformations of the objects to be recognized. The recognition is very reliable and relatively fast: the resognition time with the same slow mini-computer is about 1 sec. per object.

The realized recognition technique is a modified version of the consistent labeling. The most important modifications are the followimg:

a) The structure of an image is being described by means of a cellular complex (more precisely, block complex) rather than by a graph.

b) The "label" of an object is the full set of its topological and metric features contained in the cell list. It is geometrically exact which means that the object may be exactly reconstructed starting from the features.

c) The consistency tests are rather various and different for different object classes. They are specified in the definition of a class.

d) The definitions of the classes to be recognized are organized hierarchically: the classes of more complicated objects are defined by means of a composition of certain elementary objects which have been defined previously.

e) The definitions of the classes are being determined interactively in form of special tables specifying the necessary unary and binary relations which the features of the elementary objects must satisfy to compose an object of the class to be defined.

This technique has been also successfully used for the recognition of hand-written characters /15/

6. Conclusion

The application of the well-known notions of the topology of complexes brings many advantages for precisely and economocally describing the images, for fast and topologically precise geometrical transformations of images and for implementing flexible and reliable methods of the structural image analysis.

Literature

1. W. Rinow, Lehrbuch der Topologie, VEB Deutscher Verlag der Wissenschaften, Berlin 1975.

2. A. Rosenfeld, "Connectivity in digital pictures", J. ACM 17 (1970), 146-160.

3. T. Pavlidis, Structural Pattern Recognition, Springer-Verlag, New York 1977.

4. V.A. Kovalevsky, "Discrete topology and contour definition", Pattern Recognition Letters 2 (1984), 5, 281-288.

5. A. Rosenfeld, "Adjacency in Digital Pictures", Informat. Contr. 26 (1974), 24-33.

6. H.J. Feng and T. Pavlidis, IEEE Trans. on Circuits and Systems 22 (1975), 427-439.

7. J. Serra, Image Analysis and Mathematical Morphology, Academic Press, New York 1982.

8. H. Elliott and L. Srinivasan, "An application of dynamic programming to sequential boundary estimation", Computer Graphics and Image Processing, 17 (1981), 4, 291-314.

9. R. Klette and K. Voss, "Theretische Grundlagen der digitalen Bildverarbeitung, IV. Orientierte Nachbarschaftsstrukturen", Bild und Ton 39 (1986), 7, 213-219.

10. V.A. Kovalevsky, "Discrete geometry and image processing", Proc. Int. Conference on Computers, Systems and Signal Processing, Bangalore, India, Dec. 9-12, 1984, 934-938.

11. V.A. Kovalevsky, "On the topology of digital spaces", Proceeding of the problem-seminar "Digtale Image Processing" in Fleeth/Mirrow, GDR, April 1986, to be published by the Technical University of Dresden, 1987.

12. V.A. Kovalevsky and A. Wieczorek, "Fast Recognition of Digital Straight Segments", Technical Report 86-73-03, Institute of Cybernetics, Berlin, GDR, 1986.

13. V. Kempe, B. Rebel, W. Wilhelmi, "The Interactive Image Processing Console A6471", Proceedings of the 6th International Conference on Patternn Recognition, 1982, 607-610.

14. Nailja Luth, "Affine Transformations on Images Represented by a List Data Structure", in this issue.

15. V. Kovalevsky, "Structural image analysis", Proc. of the 8th International Conference on Pattern Recognition, Paris, 1986, 358-368.

AN AXIOMATIC APPROACH TO DISCRETE GEOMETRY AND ITS RELATIONS TO USUAL DIGITAL GEOMETRY FOR IMAGE PROCESSING

Albrecht Hübler [1)]

1. Introduction

There is a traditional and, particularly for the case of the orthogonal grid, well studied approach to define important geometric notions for grid point sets: Choose a special grid which is embedded in the Euclidean, plane, fix a digitization function, and translate notions of Euclidean geometry into geometric notions for grid point sets via digitization. Thus, these conceptions of digital geometry depend in a strong way of the considered grid and the digitization function, resulting in a variety of different "digital images" of Euclidean geometry.

During the last 15 years a lot of results was published concerning digital geometry and its relevance for theory and practice of image processing and computer graphics; the references to /2,5,6,8,9,10,11, 12/ gives only a small cut-out of literature in this field. There were produced many useful and practicable results, but also some further disadvantages has gone clear. Some fundamental properties of continuous geometry and topology do not hold in digital geometry, for example, the uniqueness of a digital straight line connecting two grid points does not hold, there are difficulties to find a reasonable definition for parallelity, the digital analogon to Jordan curve theorem is not valid in all grids /see 7/. Here the reasons can be found, why until now there is no homogeneous mathematical theory of digital geometry.

To overcome this unsatisfactory situation starting with first ideas in /3/ a system of axioms has been developed /4/, representing essential properties and relations for a geometric theory over discrete structures with relevance to image processing. The present paper is based on the approach of /4/. The theory of discrete geometry introduced in /4/ is continued and completed, its relations to traditional digital geometry become more evident. For the sake of brevity we can only present results without any proof.

[1)] Albrecht Hübler, Friedrich-Schiller-University Jena, Department of Mathematics, UHH, 17. OG, Jena, DDR-6900, German Democratic Republic

2. A theory of discrete geometry

Geometric theories which refers to our experienes from real world base on continuity assumptions. Such assumptions are invalid for objects in digital images. Here, bounded geometric objects consist of a finite number of points. Any discrete geometric theory defined by a system of axioms has to include this fact into its axioms.
In the following we present in a concise way the axioms, definitions, corollaries and properties of our abstract theory of discrete geometry. Thereby the kind of presentation is near that in /4/. Some results and notions are taken over directly from there. For the important and very interesting comparision to axiomatic theory of Euclidean geometry we refer the reader to /1/ where the rather complex axiom of Pasch is avoided, thus facilitating comparisions between that and our approaches.

Let P be a set of points and G a nonempty set of subsets of P, whose elements are called straight lines. These abstract sets P and G are basis objects of the presented theory. The axioms express essential properties and relations between basis objects, and thus represent the base for deduction of further properties and for introduction of important notions.

The first four axioms are the axioms of incidence.

<u>Axiom 1:</u> For any two different points $p,q \in P$ there is exactly one straight line $g \in G$ which contains p and q.

<u>Axiom 2:</u> For any straight line $g \in G$ there are at least three different points $p,q,r \in P$ with $p,q \in g$ and $r \notin g$.

For $p,q \in P$ with $p \neq q$ we write $g(p,q)=g(q,p)$ for the straight line containing p and q. From Axiom 2 it follows the existence of at least three points in P, and, together with Axiom 1, the existence of at least three straight lines in G. Two different straight lines can have at most one point in common. But, according to all experiences, in discrete geometry two different straight lines need not have a common point. Therefore, the usual definition of parallelity (two straight lines g_1 and g_2 are parallel if and only if $g_1 = g_2$ or $g_1 \cap g_2 = \emptyset$) is not fruitful for discrete geometry. We introduce a parallelity relation on the set G of straight lines as a third basis object by the next axiom.

<u>Axiom 3:</u> On the set G is defined an equivalence relation $\parallel$ such that for $g \in G$ and $p \in P$ there exists exactly one straight line $g' \in G$ with $g \parallel g'$ and $p \in g'$.

The set G is divided by relation into equivalence classes of pairwise parallel straight lines. We call this equivalence classes directions.

Definition 1 (Translations): A translation is a 1-1-correspondence A from P onto P which, either is the identity id ($id(p)=p$ for all $p \in P$), or fulfils the following three properties:

(T-1) For $g \in G$ is $A(g)=\{A(p): p \in g\}$ a straight line which is parallel to g.

(T-2) For all $p \in P$ is $A(p) \neq p$

(T-3) The set of straight lines $\{g(p,A(p)): p \in P\}$ is a direction.

The following axiom guarantees the existence of translations different from id.

Axiom 4: For points $p,q \in P$ there exists exactly one translation A with $A(p)=q$.

Theorem 1: Two straight lines g_1 and g_2 are parallel, if and only if, there is a translation $A \neq id$ and g_1 and g_2 are fix with respect to A, i. e. $A(g_1)=g_1$ and $A(g_2)=g_2$.

Theorem 2: Two straight lines g_1 and g_2 are parallel, if and only if, there is a translation A with $A(g_1)=g_2$.

Thus, in a concrete model of this theory it suffices to characterize the set of translations, to fix the relation in a unique way.
For the set of translations it can be shown that together with the operation $\circ$ of concatenation it forms a commutative group, i. e. for translations A,B,C hold

(i) $A \circ B$ is again a translation,

(ii) for A there is a unique inverse translation A^{-1} with $A \circ A^{-1} = A^{-1} \circ A = id$,

(iii) id is the 1-element of the group with $A \circ id = id \circ A = A$,

(iv) the operation $\circ$ is associative: $A \circ (B \circ C) = (A \circ B) \circ C$,

(v) the operation $\circ$ is commutative: $A \circ B = B \circ A$.

In the following group of axioms is fixed the essential distinction between discrete and Euclidean geometry, that in the case of discrete geometry between two points there need not to be a third one.
We use the notation order relation for a binary, transitive, irreflexive and connex (i. e. any two different elements are 'comparable') relation. Now, for all straight lines g let be given two fixed opposite order relations $<$ and $>$ on g, a fourth kind of basis objects of the presented theory. We call the pairs $[g,<]$ and $[g,>]$ oriented straight lines. For points $p,q \in g$ it holds $p < q$, if and only if, $q > p$.

We call points colinear if they belong to the same straight line. For colinear triples p,q,r of points we define a "betweenness" relation:

<u>Definition 2 (Betweenness-relation):</u> For colinear points p,q,r holds Z(p,q,r), if and only if, $p < q < r$ or $p > q > r$. In the case of Z(p,q,r) we say that the point q is between the points p and r.

<u>Definition 3 (Straight line segment):</u> Given two different points p and q, the straight line segment pq is defined by pq= r:p=r v q=r v Z(p,r,q) . The points p and q are the end points of pq and all other points of pq are called interior points of pq.

<u>Axiom 5:</u> For any oriented straight line [g,<] or [g,>] and $p \in g$ there are at least two points $q,r \in g$ with Z(q,p,r).

<u>Corollary 1:</u> For a straight line g and $p \in g$ there exists a straight line segment of g which contains p as an interior point.

<u>Axiom 6:</u> For points $p,q \in P$ there exist at most a finite number of points which are between p and q.

This Axiom 6 is the important axiom of discretness. From this moment the Euclidean geometry is excluded of the class of models of our theory.

<u>Corollary 2:</u> Straight line segments are finite sets, containing at least the two end points.

<u>Theorem 3:</u> Any straight line contains exactly enumerable infinitely much points.

A consequence of theorem 3 is the fact that any oriented straight line can be represented as point sequence $\ldots\ldots p_{-2}, p_{-1}, p_0, p_1, p_2, \ldots\ldots$ where $q_i < q_{i+1}$ holds for all indices i, and all points of g occure in this sequence.

<u>Definition 4 (Half line):</u> Given on oriented straight line [g,<] and a point $p \in g$, the sets $\{q: q \in g \;\&\; (q=p \vee q<p)\}$ and $\{q: q \in g \;\&\; (q=p \vee q > p)\}$ are the half lines on g with initial point p.

From the possibility of representation of straight lines as ordered point sequences it follows, that a half line is well defined by its initial point p and another of its points, say q. Let pq^+ denote this half line, and pq^- denote the opposite half line with initial point p, i. e. $pq^+ \cap pq^- = \{p\}$ and $pq^+ \cup pq^- = g(p,q)$.

The fundamental relation of translational equivalence is defined for pairs N and M of point sets: We call M and N translational equivalent, if and only if, there is a translation A with $A(M) = N$.
On the set of all half lines the relation of translational equivalence induces a decomposition into equivalence classes which we call direction sense. Thus any direction corresponds with exactly two direction senses.

<u>Axiom 7:</u> Let g_1, g_2, g_3 be pairwise distinct and pairwise parallel straight lines, and p_i, q_i points of g_i for $i = 1,2,3$. Then, from $Z(p_1, p_2, p_3)$ and from the colinearity of q_1, q_2 and q_3, it follows $Z(q_1, q_2, q_3)$.

The idea of Axiom 7 meets the expectation that parallel straight lines do not "cross" each other: g_2 cannot break out of the "stripe" defined by g_1 and g_3. The following theorem guarantees, that for any two parallel straight lines g_1 and g_2 the premise of Axiom 7 can be realized.

<u>Theorem 4:</u> For two parallel straight lines g_1 and g_2 there is a straight line g_3 which is parallel to g_1 and g_2, such that there are infinitely many triples of colinear points p_1^j, p_2^j, p_3^j, $j = 0,1,2,3,\ldots$ with $p_i^j \in g_i$ for $i = 1,2,3$.

Using Axiom 7 the invariance of betweenness-relation under translations can be proved:

<u>Theorem 5:</u> For any three points p,q,r with $Z(p,q,r)$ and an arbitrary translation A it follows $Z(A(p), A(q), A(r))$.

The introduction of the very useful notion of simple translations is important to derive a very instructive connection between straight lines and translations.

<u>Definition 5 (Simple translation):</u> A translation A is called simple, if and only if, for all translations $B \neq A$ there is no integer n with $A = B^n$.

Thereby B^n denotes the n times repeated concatenation of B. In other words, a translation A is a simple one, if it cannot be generated by repeated application of another translation B.

<u>Theorem 6:</u> For any straight line g there is a suitable simple translation A, such that for arbitrary $p \in g$ the straight line g can be repre s ented in the form $g = \{A^n(p): n \in \Gamma\}$, where Γ is the set of all

integers. (We call this representation the standard form of g.)

This theorem gives interesting and essential insights into the structure of models of the presented theory. Its content will be completed by two corollaries:

Corollary 3: For a straight line g there are exactly two simple translations A and A^{-1} which allow a representation of g in standard form. (We call A and A^{-1} the generators of g.)

Corollary 4: Parallel straight lines have the same generators.

The next theorem gives an analog property to Axiom 7 for nonparallel straight lines with a common point:

Theorem 7: Let g_1, g_2, g_3 be pairwise different straight lines with a common point p. For points $p_i \in g_i$, i = 1,2,3, let be $Z(p_1,p_2,p_3)$. Let further A_i denote the translations defined by $A_i(p)=p_i$, and $p_i' = A_i^{-1}(p)$, i = 1,2,3. Then also $Z(p_1',p_2',p_3')$ is valid.

The last axiom introduced in this paper is directed to the existence of at least two straight lines with the property to cover the whole point set P when translating one along the other.

Axiom 8: There are two straight lines g_1 and g_2, such that for any point $p \in P$ there exists a point $q \in g_1$ with $g(p,q) \parallel g_2$.

It is easy to see that P also is covered, if vice versa, g_1 is translated along g_2.

Corollary 5: Let straight lines g_1 and g_2 fulfil Axiom 8. Then this is true for any pair g_1' and g_2' of straight lines with $g_1' \parallel g_1$ and $g_2' \parallel g_2$.

Now it is possible to fit out our geometry with coordinate systems: Let $g_x = \{A_x^n(s) : n \in \Gamma\}$ and $g_y = \{A_y^n(s) : n \in \Gamma\}$ be two straight lines in standard form which suffice Axiom 8, and let s be their common point. Obviously, for any point $p \in P$ there exist a unique pair of integers i, j with $p = A_x^i \circ A_y^j(s)$. Thus, a 1-1-correspondence between points and all pairs of integers is fixed. We call i the x-coordinate and j the y-coordinate of p, respectively, and write in general, for brevity, x(p) and y(p) to denote the two coordinates of a point p, with respect to a fixed coordinate system with x-axis g_x and y-axis g_y.

Definition 6 (Orientation): An ordered triple (p_1,p_2,p_3) of pairwise different points is positive oriented (negative oriented), if and only if, the value of the determinant

$$\begin{vmatrix} 1 & x(p_1) & y(p_1) \\ 1 & x(p_2) & y(p_2) \\ 1 & x(p_3) & y(p_3) \end{vmatrix} = \mathcal{O}r(p_1,p_2,p_3)$$

is positive (is negative).

<u>Theorem 8:</u> For points p_1,p_2,p_3 the identity $\mathcal{O}r(p_1,p_2,p_3) = 0$ is valid, if and only if, p_1,p_2 and p_3 are colinear.

From theorem 8 we know, that for two fixed points p_1 and p_2 and for any point q which not belongs to $g(p_1,p_2)$ the triple (p_1,p_2,q) either is positive oriented or is negative oriented. Moreover, it is possible to show:

<u>Theorem 9:</u> Let be given a straight line $g = \{A^n(p): n \in \Gamma\}$ in standard form. Further, let be $p_1,p_2,p_1',p_2' \in g$ and $q \notin g$, where for $p_2 = A^i(p_1)$ and $p_2' = A^j(p_1')$, either $0 < i,j$, or $0 > i,j$ (i. e. the half lines $p_1p_2^+$ and $p_1'p_2'^+$ have the same direction sense). Then it holds $\mathcal{O}r(p_1,p_2,q) > 0$, if and only if, $\mathcal{O}r(p_1',p_2',q) > 0$.

<u>Corollary 6:</u> An arbitrary straight line $g = g(p_1,p_2)$ induces, with respect to the direction sense of $p_1p_2^+$, a decomposition of the set of all points not belonging to g into exactly two disjoint classes

$H^+(g) = \{q: q \notin g \ \& \ \mathcal{O}r(p_1,p_2,q) > 0\}$ and

$H^-(g) = \{q: q \notin g \ \& \ \mathcal{O}r(p_1,p_2,q) < 0\}$, called the two

halfplanes according to g.

<u>Definition 7:</u> (Convexity): A subset M of P is convex, if and only if, for any two different points p and q of M the straight line segment pq is a subset of M.

Some important and reasonable properties of halfplanes are expressed in the next theorem.

<u>Theorem 10:</u> The halfplanes $H^+(g)$ and $H^-(g)$ induced by a straight line g fulfil the following statements:

(a) $H^+(g)$ and $H^-(g)$ are maximal convex subsets of P-g.

(b) For any straight line g' holds,
either $g = g'$, or $g' \subseteq H^+(g)$, or $g' \subseteq H^-(g)$, or there are points $p,q \in g'$ with: $pq^+ \subseteq H^+(g)$ and $pq^- - \{p\} \subseteq H^-(g)$, or $pq^+ - \{p\} \subseteq H^+(g)$ and $pq^- - \{p\} \subseteq H^-(g)$.

(c) For any straight line $g' \neq g$ the relation $g \parallel g'$ is valid, if and only if, $g' \subseteq H^+(g)$ or $g' \subseteq H^-(g)$.

Definition 8 (Strong convexity): A subset M of P is strong convex, if and only if, M can be represented as intersection of a finite number of halfplanes.

We will see later, in an example, that not all convex sets are strong convex sets.

Before we will consider models of this theory of discrete geometry we resume the considerations of this section by emphasizing the central role of translations: It is sufficient to know the set of translations on P, to deduce all further notions of our geometry. Indeed, starting with the translations on P, the set of straight lines G, the parallelity relation and the order relations on the straight lines are determined in a unique way, i. e. all basis objects are known.

3. The models of theory of discrete geometry

In conformity with /13,14,15/ we consider neighbourhood graphs as a suitable mathematical description of carrier structures of digital images. A neighbourhood graph GR(P,N) is the graph representation of a symmetric and irreflexive relation **N** (the neighbourhood relation) on a nonempty set P, i. e. the nodes of GR(P,N) are the elements of P, where $p,q \in P$ are connected by an undirected edge, if and only if, $(p,q) \in \mathbf{N}$. Let d_{GR} denote the metric on P induced by GR(P,N), which defines the distance between two nodes as the length of a shortest connecting path in GR(P,N).

We now from section 2 that any model of theory of discrete geometry can be completely characterized by specifying its point set and the set of translations.

Definition 9 (Translation on GR(N,P)): A bijective mapping A from P onto P is called a translation on GR(N,P), if an only if, for all $p,q \in P$ hold

(t1) $(p,q) \in N \longleftrightarrow (A(p),A(q)) \in N$ (A is an automorphism on GR(P,N))

(t2) $d_{GR}(p,A(p)) = d_{GR}(q,A(q))$ (a constant translation distance)

On our way to specify models of the theory of discrete geometry we have to exclude those neighbourhood graphs which do not fulfil some fundamental properties for their set T_{GR} of translations:

(p-1) The set T_{GR} is closed under concatenation $\circ$.

(p-2) The operation $\circ$ is commutative on T_{GR}.

(p-3) For $p,q \in P$ with $(p,q) \in N$ there is a translation $A \in T_{GR}$ with $A(p)=q$.

(p-4) For all translation $A \in T_{GR}$ with $A \neq id$ and arbitrary $p \in P$ there is no integer n with $A^n(p)=p$.

Remark, that properties (p-1),(p-2),(p-3), and (p-4) are necessary conditions for GR(P,N) to be a model of our theory. Therefore, from now we consider only those neighbourhood graphs GR(P,N) whose translation set T_{GR} realizes (p-1),(p-2),(p-3), and (p-4); we call them neighbourhood graphs with honest translation set.

Corollary 7: All neighbourhood graphs GR(P,N) with honest translation set, and with more than one node, fulfil the following properties (a), (b), (c), (d), (e):

(a) P is an enumerable ifinite set.

(b) Any $p \in P$ has the same constant number of neighbours.

(c) For $p,q \in P$ there is a translation $A \in T_{GR}$ with $A(p) = q$.

(d) For $p \in P$ and $A_1, A_2 \in T_{GR}$ from $A_1(p)=A_2(p)$ the identity $A_1=A_2$ follows.

(e) The set T_{GR} is an enumerable infinite set.

We call translations of T_{GR} with translation distance 1 elementary translations.

Corollary 8: Let ν be the (for all $p \in P$ constant) number of neighbours of the nodes of a neighbourhood graph GR(P,N) with honest translation set. Then there are exactly ν elementary translations in T_{GR}.

Theorem 11: For a neighbourhood graph GR(P,N) with honest translation set the set E_{GR} of elementary translations is a generating system of T_{GR}, i. e. any $A \in T_{GR}$ can be represented as a finite linear combination of elementary translations.

With theorem 11 we have for neighbourhood graphs with honest translation set the existence of finite generating systems for T_{GR}. We call the linear independent finite generating systems of T_{GR} basis of T_{GR}, and define the dimension of GR(P,N) to be the cardinality of a smallest basis of T_{GR}. Common examples for 2-dimensional neighbourhood graphs with honest translation set are the so called 4-, 6-, and 8-neighbourhood graphs (see fig. 1) The 3-neighbourhood graph (fig. 1) not realizes property (p-3), and therefore has no honest translation set.

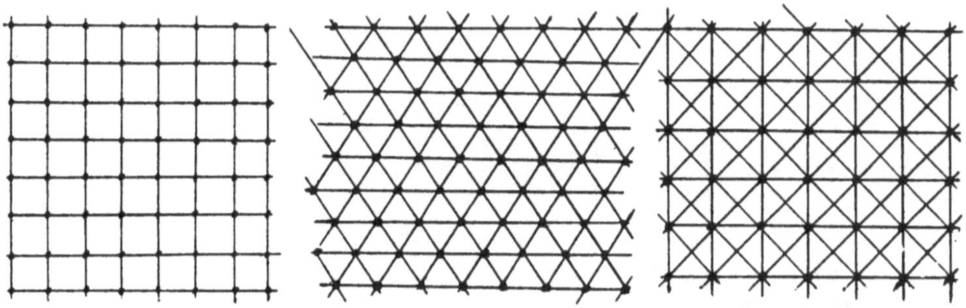

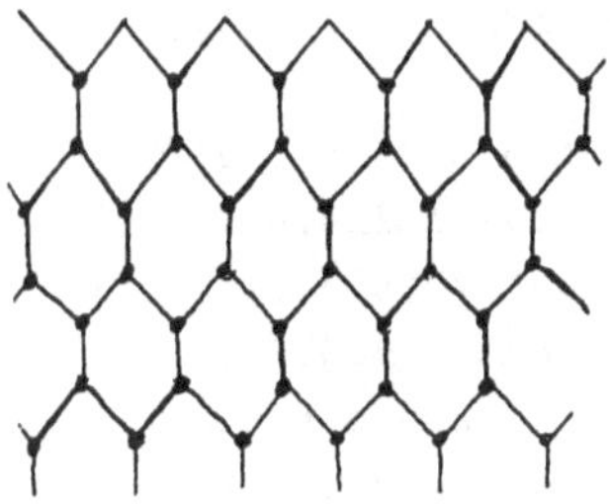

fig. 1: 4-, 6-, 8-neighbourhood graphs are neighbourhood graphs with honest translation set.
The 3-neighbourhood graph allows no translations with an odd translation distance.

The next pair of theorems describes the class of models of theory of discrete geometry.

Theorem 12: Let GR(P,N) be a 2-dimensional neighbourhood graph with honest translation set T_{GR}. Then the pair $[P, T_{GR}]$ is a model of the theory of discrete geometry.

In other words theorem 12 means that considering P as point set, and defining the other basis objects: the set G of straight lines, the parallelity relation $\parallel$, and the order relations on straight lines, in conformity with the translation set T_{GR}, the axioms 1 to 8 will be fulfilled. With theorem 12 we know a class of models which includes the discrete geometries of 4-neighbourhood graph, of 6-neighbourhood graph, and of 8-neighbourhood graph.

Definition 10 (Isomorphic models): Let be $[P1,T1]$ and $[P2,T2]$ two models of theory of discrete geometry, defined by the point sets P1 and P2 and the translation sets T1 and T2, respectively. We call $[P1,T1]$ and $[P2,T2]$ isomorphic, if and only if, there is a bijection I from P1 onto P2 which fits the condition of translation invariance:
Let $p_1, q_1, r_1, s_1 \in P1$ and $p_2=I(p_1)$, $q_2=I(q_1)$, $r_2=I(r_1)$, $s_2=I(s_1)$, where for the translation $A_1 \in T1$ hold $A_1(p_1)=q_1$ and $A_1(r_1)=s_1$. Then for the translation $A_2 \in T_2$ defined by $A_2(p_2)=q_2$ it follows the identity

$A_2(r_2)=s_2$.

Theorem 13: Let $GR^4(P^4,N^4)$ be the 4-neighbourhood graph. Then any model $[P,T]$ of the theory of discrete geometry is isomorphic to the discrete geometry $[P^4,T_{GR^4}]$ of 4-neighbourhood graph.

According to this theorem the 4-neighbourhood graph can be used as a standard model of theory of discrete geometry. This fact emphasizes the significance of comparisions with traditional digital geometry in the orthogonal grid.

4. Straight lines and convexity in discrete and digital geometry

The notions of straight lines and convexity are the most studied objects in digital geometry. There are known several different definitions, characterizations and connections (see /2,5,6,9,12/ for example), most of them with respect to the orthogonal grid.
We will consider the notions of digital straight line and digital convexity with respect to the planar orthogonal grid and to the standard grid intersection digitization scheme (see /9/): A set S of grid points is a digital straight line, if and only if, there is a (real) straight line whose digital image, with respect to standard grid intersection digitization scheme, is S (see /2,9/ for comparison). A set M of grid points is digital convex, if and only if, there is a (real) convex set whose digital image, with respect to standard grid intersection digitization scheme, is M (compare /2,5/).

For any set M of grid points the border b(M) is the set of all points $p \in M$ which have a 4-neighbour q not belonging to M.

The following theorems concerning relations between notions of straight line and convexity in digital and discrete geometry are of practical and theoretical interest. They are illustrated in figure 2.

Theorem 14: The border b(H) of an arbitrary (discrete) halfplane is a digital straight line. If S is a digital straight line with rational slope (i. e. its real preimages have rational slope) then there is a (discrete) halfplane H such that S = b(H).

We say that a discrete straight line g corresponds to a digital straight line S, and vice versa, if and only if, S is the border of one of the halfplanes induced by g.

Corollary 9: Any discrete straight line corresponds to exactly two digital straight lines. Any digital straight line with rational slope corresponds to exactly two discrete straight lines.

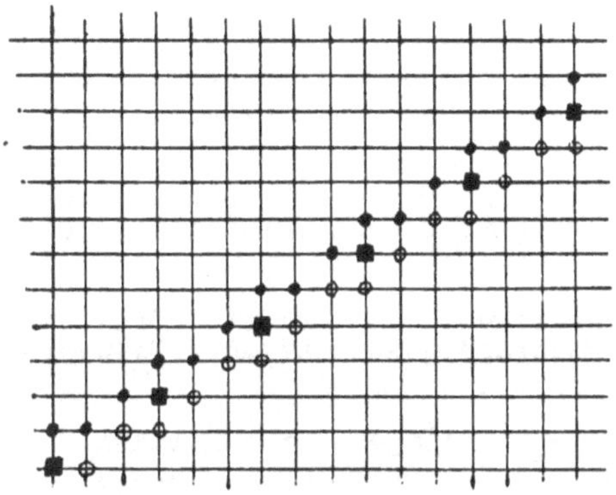

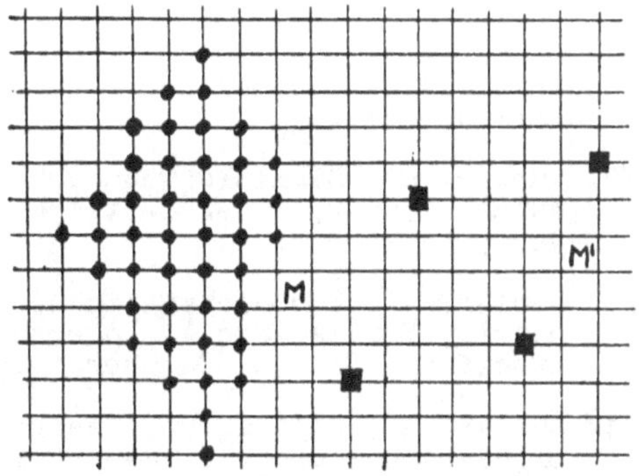

A discrete straight line▪and its two corresponding digital straight lines: o •

A discrete strong convex set M, and a discrete convex set M' which is not digital convex.

fig. 2

Theorem 15: Let M b a connected grid point set. Then M is digital convex, if and only if, M is (discrete) strong convex. Any digital convex set is also discrete convex.

5. Conclusions

It is presented an axiomatic theory of discrete geometry, which is selfcontained, and which is defined in an abstract way without any respect to concrete discrete structures. This theory has models, i. e. there are concrete interpretations of basis objects, which are of interest from the point of view of digital image processing. The usual 4-neighbourhood graph (the planar orthogonal 4-grid) can be considered as a standard model of this theory. Results of section 2 show that the presented theory allows faithful definitions of important geometric notions together with reasonable properties and connections between them. The interesting relations between fundamental geometric notions like straight lines and convexity in digital geometry on the one hand, and in the discrete geometry on the other hand, encourage the author to continue the investigations for extension of the theory of discrete geometry and for a deeper study of its relevance as a theoretical mathematical instrument for solving geometric problems of image processing and computer graphics.

6. References

/1/ Böhm, J., et.al., Geometrie. I. Axiomatischer Aufbau der euklidischen Geometrie, Deutscher Verlag der Wissenschaften, Berlin, 1979.

/2/ Hübler, A., Digitale Geometrie - Einführung und Algorithmen, in AUTBILD '87/2, Wiss. Beiträge der Friedrich-Schiller-Universität Jena, to appear.

/3/ Hübler, A., Zur mathematischen Grundlegung der diskreten Geometrie, Proceedings 1. Internat. Fachtagung Automatische Bildverarbeitung, Berlin, Oktober 1985.

/4/ Hübler, A., Klette, R., Theoretische Grundlagen der digitalen Bildverarbeitung VII. Geometrie auf Nachbarschaftsstrukturen, Bild und Ton 40 (1987), to appear.

/5/ Kim, C. E., Rosenfeld, A., Digital straight lines and convexity of digital regions, IEEE Trans. Pattern Anal. Mach. Intell. PAMI-4, 1982, 149-153.

/6/ Klette, R., The m-dimensional grid point space, Computer Vision, Graphics and Image Processing 30, 1985, 1-12.

/7/ Kovalevsky, V. A., Discrete topology and contour definition, Pattern Recognition Letters 2 (1984) 281-288.

/8/ Kovalevsky, V. A., Selfcontained foundations of discrete geometry, in GEOBILD '85, Proceedings Internat. Symposium Geometric Problems of Image Processing, Georgenthal, 1985, 49-57.

/9/ Rosenfeld, A., Digital straight line segments, IEEE Trans. Comput. C-23, 1974, 1264-1269.

/10/ Rosenfeld, A., Kak, A. C., Digital Picture Processing (2nd ed.), Academic Press, New York, 1982.

/11/ Serra, J., Image Analysis and Mathematical Morphology, Academic Press, London, 1982.

/12/ Sklansky, J., Recognition of convex blobs, Pattern Recognition 2 (1970), 3-10.

/13/ Voss, K., Hufnagl, P., Klette, R., Theoretische Grundlagen der digitalen Bildverarbeitung I. Einleitung, Bild und Ton 38 (1985), 10, 299-310.

/14/ Voss, K., Hufnagl, P., Klette, R., Theoretische Grundlagen der digitalen Bildverarbeitung II. Nachbarschaftsstrukturen, Bild und Ton 38 (1985), 11, 325-331.

/15/ Voss, K., Hufnagl, P., Klette, R., Theoretische Grundlagen der digitalen Bildverarbeitung IV. Orientierte Nachbarschaftsstrukturen, Bild und Ton 39 (1986), 7, 213-219.

TWODIMENSIONAL TOPOLOGIC ORDERS AND PICTURE PROCESSING

H. Kiesewetter

ACADEMY OF SCIENCES OF THE GDR,
CENTRAL INSITUTE FOR CYBERNETICS AND INFORMATION PROCESSES,
Kurstrasse 33, Berlin, GDR - 1086

1. Topologies of the raster

Objects of picture processing are considered to be distributions G(X,Y) on a raster

$$X = 0 , 1 , .. , L \quad (1)$$
$$Y = 0 , 1 , .. , L$$

The elements (X,Y) of the raster are called pixels.
In order to get a topology of the raster we have to define the contacts of each element with its direct neighbours. The contacts are established in clockwise order around each element in form of a star. Several types of stars have been considered (/ 4 /, cf. Fig. 1).
Topologies of the raster are examples of a special type of topologies called integer topologies. There are remarkable differences as well as analogies between integer topologies and other topologic concepts.
Topology is governed by two basic operations: Connexion and separation.
If arbitrary two elements of a subset are connected by a chain of consecutive contacts between direct neighbours, the subset is called a continuum. A continuum is a connected component in the sense of graph theory. It contains at least two elements. A subset consisting of one element only will not be called a continuum.
An element is called boundary element, if it contains at least one direct neighbour which does and another direct neighbour which doesn't belong to the continuum. The boundary of a continuum, which is a proper subset of the raster, has to be divided into two parts. The inner boundary contains all boundary elements, which itself do belong to the continuum, and the outer boundary all boundary elements, which itself don't belong to the continuum.

Each chain connecting an element of the continuum with an element outside contains at least one element of the inner and one element of the outer boundary. Therefore, the inner boundary as well as the outer boundary, each for itself alone, separate a continuum from its complement.

4 - topology

P _ (1 , 3 , 5 , 7)

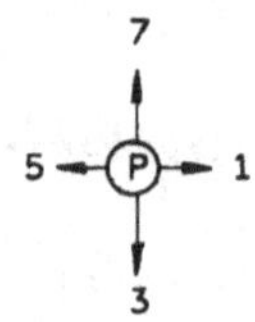

6M - topology
(main diagonal)

P _ (1 , 2 , 3 , 5 , 6 , 7)

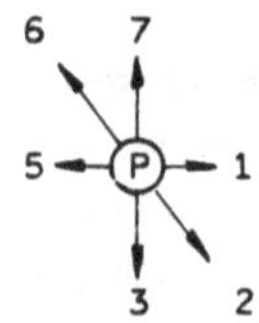

6S - topology
(secondary diagonal)

P _ (7 , 8 , 1 , 3 , 4 , 5)

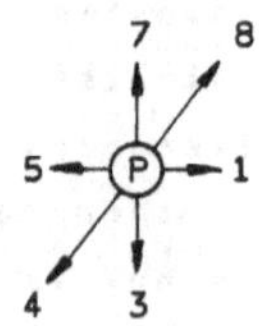

8 - topology

P _ (1 , 2 , 3 , 4 , 5 , 6 , 7 , 8)

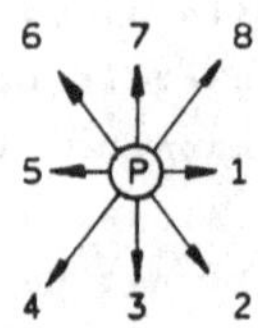

Fig. 1. Topologies of the raster

For integer topologies it has to be noticed, that the topology of a continuum and the topology of its inner resp. outer boundary may be different. For the above mentioned topologies the following couplings have been established (/ 4 /):

continuum	boundary
4 - topology	8 - topology
8 - topology	4 - topology
6M - topology	6M - topology
6S - topology	6S - topology

2. Axioms of topologic order

Topologies of the raster serve as a guideline for the axiomatic definition of integer topologies. We start with the definition of positional order.

Let PK be a finite set of elements K, called contacts and PS a set of subsets S of PK, called stars. We state the following axioms of positional order:

(P1) For every contact K exists a uniquely determined inverse contact -K.

(P2) The inverse of the inverse contact is the contact itself.

(P3) A contact and its inverse are two distinct elements.

(P4) If contact K1 differs from contact K2, then K1 is either prior or (exclusively) posterior to K2.

(P5) Every contact is an element of exactly one star.

(P6) If star S1 differs from star S2, then each contact K1 of S1 is either prior or (exclusively) posterior to all contacts K2 of S2, i.e. S1 is either prior or (exclusively) posterior to S2.

(P7) If PI(S) designates the first contact of the star S, called pivot of S, then -PI(S) is prior to PI(S) for each star S, posterior to the first star.

(P8) The inverse -PI(S) of the pivot PI(S) of S is prior to the inverses of all other elements of S for each star S, posterior to the first star.

Remarks to the axioms:

Axiom (P1) implies that each contact occurs twice in the order, namely in connexion with its INIT and EXIT. The following relations hold for all K of PK:

$$-(-K) = K \qquad (2)$$

$$-K \neq K \qquad (3)$$

Axiom (P4) establishes an order relation on PK. According

to axiom (P5), PS may be considered to be a covering of PK by means of non-intersecting subsets. Furthermore, as it is stated by axiom (P6), each star consists of an intervall of consecutive contacts with respect to the order relation on PK.

The existence of a first contact in each star and a first star in PS follows from the fact, that all sets are finite. Because of axiom (P4) all first elements are uniquely determined.

The first contact of the first star plays a special role. It can be arbitrarily chosen for each positional order. It will be called pivot of the positional order and designated by PI(P). We have

$$PI(P) < K \qquad (4)$$

for all K different from PI(P) and especially

$$PI(P) < -PI(P) \qquad (4')$$

Axiom (P7) as well as (P8) are valid for all pivots PI(S) different from PI(P). Axiom (P7) takes the form

$$-PI(S) < PI(S) \qquad (5)$$

which is just the opposite relation to (4') valid for PI(P). It means that the contact -PI(S) serves as a link between the star S and a prior one. Therefore, each positional order is a continuum.

From Axiom (P8) we get the relation

$$-PI(S) < -K \qquad (6)$$

for all contacts K of S different from PI(S). Thus the pivot of a star, posterior to the first star, is determined by the first occuring inverse contact.

Three types of contacts have to be distinguished in a star: The inverse of a contact can be a contact of a star prior, equal or posterior to the corresponding star. A pivot PI(S) (except PI(P)) belongs to the first type.

Definition 1:

> An integer topology is a positional order or a sequential union of positional orders.

3. Algebraic - topologic operations

Representation of topologies by cyclic orders of neighbours and edges of opposite sign have successfully been used by

Speck (/ 5 /). His work will be continued by introducing algebraic-topologic operations (/ 1 /, / 2 /).

Let

$$PKP = \{ K(1), K(2) \ldots \} \quad (7)$$

$$PKN = \{ -K(1), -K(2) \ldots \} \quad (7')$$

be the positive and negative part of PK. We define a star of contacts by the algebraic-topologic formulas

$$(K(1), K(2) \ldots K(V)) \quad (8)$$

$$INIT (K(1)) = INIT (K(V)) \quad (8')$$

$$INIT (K(I)) = INIT (K(I-1)) \quad (8")$$

for I=2,3,,, V. The common INIT of the contacts will be called a position P. The contacts are assumed to be arranged in clockwise order around the position.

A positional order, according to definition 1, can be represented as a sequential union of stars fulfilling the axioms (P1) - (P8).

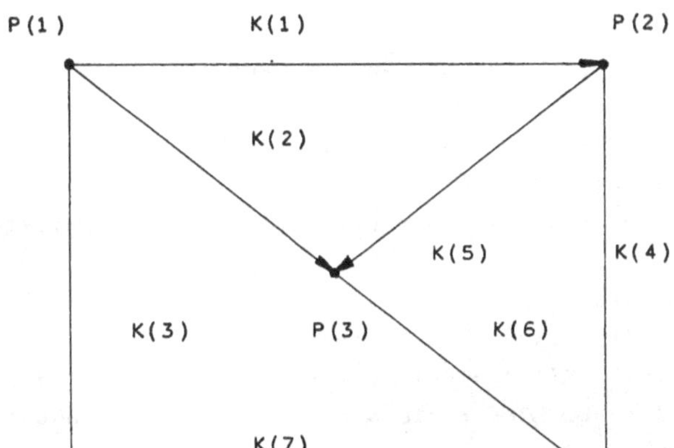

Fig. 2. Integer topology

Example 1:

(K(1), K(2), K(3)).

(-K(1), K(4), K(5)).

(-K(2), -K(5), K(6)).

(-K(3), K(7)).

(-K(4), -K(7), -K(6))

Each positional order may be transformed into a regional

order by means of the regionalization algorithm.

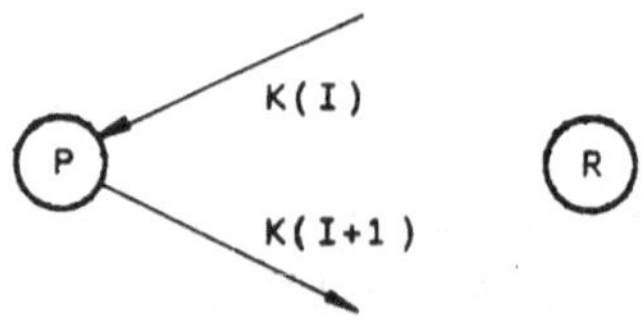

Fig. 3. Regionalization algorithm

At each step of regionalization an angle

P: (... -K(I) . K(I+1) ...)

with respect to a position P is transformed into an angle

R: (... K(I) . K(I+1) ...)

with respect to the corresponding region R. Clockwise order around positions produces anti-clockwise order around regions.

Example 2:

(K(1) . K(4) . -K(7) . -K(3)) ,
(-K(1) . K(2) . -K(5)) ,
(-K(4) . K(5) . K(6)) ,
(K(7) . -K(6) . -K(2) . K(3))

We define a cycle of contacts by the algebraic-topologic formulas

(K(1) . K(2) ... K(U)) (9)
INIT (K(1)) = EXIT (K(U)) (9')
INIT (K(I)) = EXIT (K(I-1)) (9")

for I=2,3,., U. The cycle constitutes the boundary of a region R.

The axioms of positional order (P1) - (P8) become the axioms of regional order (R1) - (R8) by a slight modification of the notation: Cycle instead of star, RK instead of PK, RC instead of PS and C instead of S. Positional and regional order may be considered to be dual versions of one consistent concept of topologic order.

Definition 2:

An integer topology is a regional order or a parallel union of regional orders.

Definition 2 is not in contradiction to definition 1, because the same integer topology can be defined in each of its dual forms.

Each regional order can be transformed into a positional order by means of the positionalization algorithm, which is just the dual algorithm of the regionalization algorithm.

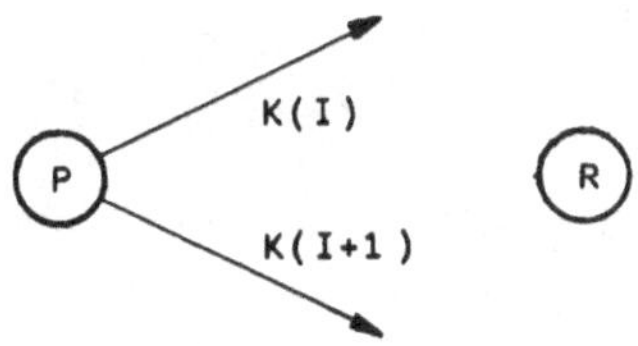

Fig. 4. Positionalization algorithm

At each step of positionalization an angle

R: (... -K(I) . K(I+1) ...)

with respect to a region R is transformed into an angle

P: (... K(I) , K(I+1) ...)

with respect to the corresponding position P.

The definition of positional and regional order is based on the identification of contacts. It may equally be based on the identification of positions or regions.

The formula

$$P1 \_ P2 \qquad (10)$$

defines a contact as a connexion of two positions P1 and P2 and induces a longitudinal orientation of the contact with the following rule of sign:

$P2 > 0$	P2 EXIT - neighbour	(10')
$P2 < 0$	P2 INIT - neighbour	(10")

The formula

$$R1 \; , \; R2 \qquad (11)$$

defines a contact as a seperation between two regions R1 and R2 and induces a transversal orientation of the contact with the following rule of sign:

$R2 > 0$	R2 RIGHT - bank	(11')
$R2 < 0$	R2 LEFT - bank	(11")

Applying these formulas we get four further forms of a topologic order, which will be demonstrated by means of the example.

Example 3:

(P(1) _ (P(2) , P(3) , P(4))) .
(P(2) _ (-P(1) , P(5) , P(3))) .
(P(3) _ (-P(1) , -P(2) , P(5))) .
(P(4) _ (-P(1) , P(5))) .
(P(5) _ (-P(2) , -P(4) , -P(3)))

Example 4:

(P(1) _ P(2) _ P(5) _ -P(4) _ -P(1)) ,
(P(2) _ -P(1) _ P(3) _ -P(2)) ,
(P(5) _ -P(2) _ P(3) _ P(5)) ,
(P(4) _ P(5) _ -P(3) _ -P(1) _ P(4))

Example 5:

(R(1) , R(2) , R(4) , R(1)) .
(R(2) , -R(1) , R(3) , R(2)) .
(R(4) , -R(2) , -R(3) , R(4)) .
(R(1) , -R(4) , R(1)) .
(R(3) , -R(1) , -R(4) , -R(3))

Example 6:

(R(1) , (R(2) . R(3) . -R(4) . -R(4))) ,
(R(2) , (-R(1) . R(4) . -R(3))) ,
(R(3) , (-R(1) . R(2) . R(4))) ,
(R(4) , (R(1) . -R(3) . -R(2) . R(1)))

Integer topology and set-theoretic (analytic) topology in the sense of Hausdorff (/ 3 /) are based on different concepts.

A position and the star of its direct neighbours form a neighbourhood of integer topology which cannot be split into proper parts. The intersection of two such neighbourhoods will not be a neighbourhood. But this is in contradiction with a basic property of analytic topologies.

The distinction between inner and outer boundary is beyond the concept of analytic topology.

There are more analogies between integer topology and algebraic topology (/ 3 /). The elements of an integer topology may be arranged in such a way that they form the cells of a twodimensional complex of algebraic topology. However, dealing with incidence relations it has to be stated, that all incidences between the cells of a complex are included into the definition, while integer topology deduces dual incidences from topologic order.

4. Local and global topology of a picture

The local topology of a picture has to be defined as a sub-topology of a topology of the full raster. Only contour pixels are taken as positions of the local topology. Since 8-topology isn't planar, but the local topology should be planar, the contacts betweeen contour pixels have to be cleared in such a way that planarity condition holds.
As compared with the topology of a full raster, there is a remarkable reduction in the number of positions (pixels) due to the fact, that inner pixels of the regions are no elements of the local topology.
Further reduction is achieved, when going over to the global topology. Contour pixels of valency 2 are connected to sections. Each section is taken as a direct connexion between positions of the global topology. Therefore, only pixels of valency 1, 3, 4, 5, 6, 7 and 8 become positions of the global topology.
The chain codes of the sections represent just the metric of the picture. This may also be given by lists of coordinates of positions in the framework of some interpolation schema.

Literature:

/ 1 / Kiesewetter, H.: Beitraege zur Digitalgraphik I, Algebraische Strukturen in der Digitalgraphik, Wiss. Zeitschrift Univ. Rostock, 21 (1972), 707-714

/ 2 / Kiesewetter, H.: Product definitions and data models for integrated CAD/CAM systems , Advances in informational aspects of industrial automation, CICIP, Berlin (1985), 180-194

/ 3 / Lefschetz, S.: Algebraic topology, Amer. Math. Soc., Coll. Publ., Vol. XXVII (1942)

/ 4 / Rosenfeld, A., Kak, A.C.: Digital picture processing, Academic Press, New York (1982)

/ 5 / Speck, P.T.: Kombinatorische Repraesentation von Bildgraphen, Mustererkennung (1984), 319-325

ON SOME BASIC CONCEPTS OF TEXTURE ANALYSIS

Dmitry Chetverikov

Computer and Automation Institute,
Hungarian Academy of Sciences,
Budapest, POB 63.
H-1502 Hungary

Abstract: Self-similarity is the generic property of the class of homogenous image textures. The human-centered approach and the operational approach to self-similarity are discussed. In the framework of the operational approach, an attempt is made to clarify some basic notions of texture analysis. The concept of texture with respect to a given similarity relation is introduced. The minimum patch constraint is an important feature of the concept. Two segmentation problems related to texture are considered. In texture mosaic segmentation, the minimum patch constraint leads to the intrinsic ambiguity of the mosaic borders. Local texture imperfections are also viewed as ambiguity zones or discontinuity regions which locally break the homogeneity of a pattern.

Key words: image texture, texture segmentation, texture imperfections.

Introduction.

Comparing different ways in which the scientists understand the term texture one finds a common feature which may be called statistical self-similarity. All patches of a homogenous texture must be similar in a sense. This seems to be the generic property of the class of homogenuos image textures. The principal question then concerns the nature of that similarity of texture patterns. What are the features to be compared ? Are there any "absolute" features whose computation would result in a "complete description" of a texture ?

Basically, there are two alternative attitudes toward the texture similarity. We call them the human-centered and the operational attitudes [1]. The human-centered approach is that two texture patches are considered similar if they are perceived as similar by a human observer (see, for example, [2]). This approach involves a subjective human decision upon the similarity of texture patterns. A human observer often judges two texture patches to be identical or not using his/her individual knowledge of their physical nature (say, two pieces of bark of tree). In other cases, comparison criteria are assumed, sometimes only implicitely, before the comparison is done. The human-centered approach usually implies the existence of perceptually motivated "absolute" texture features.

The operational attitude is the alternative to the human-centered one. The essence of the operational approach is to specify those features which are relevant in a particular texture analysis problem, and to define a similarity criterion using these features. This approach implies no "absolute" features and uses formalized decision rules.

We view digital texture as a discrete field subject to measurement with the purpose of analysing certain properties of the field. In the great majority of application problems, the criteria of what should be considered similar are either given a priory, or specified by a supervisor in the course of a preliminary analysis of the imagery in question. We shall speak of texture with respect to a given set of features paying special attention to the generic property of the class of homogenous textures, the self-similarity.

The purpose of this paper is to clarify, in the framework of the operational approach, some basic notions of texture analysis and to give formalized statements of two problems related to image segmentation using texture.

In Section 1, we introduce the notion of texture with respect to a similarity relation. The minimum patch constraint which originates from the statistical character of texture is included in the definition.

Section 2 deals with texture mosaic segmentation. Because of the minimum patch constraint, mosaic borders give rise to the ambiguity zones which are inherent in texture segmentation. A new formalized statement of the mosaic segmentation problem is given that takes into account the presence of the ambiguity zones.

In automatic surface inspection, it is often necessary to detect small surface defects which appear as local non-textured imperfections embedded in a homogenous texture. This task is clearly different from the traditional segmentation by texture. In Section 3, we introduce the formalized concept of texture imperfection and discuss some questions related to defect detection in textures.

1. Homogenous textures.

Self-similarity is the property that distinguishes the class of homogenous image textures from any other class of images. The self-similarity means that all the sufficiently large subimages of a texture image are considered similar independently of their position, shape, and size.

Textural properties can be associated only with those subimages whose size is large enough to allow for the definition of the necessary features. This constraint will be called the minimum patch constraint. Once the minimum subimage size is established, all patches of a homogenous texture pattern are expected to give similar values of the features.

Consider a digital image X and the set X_D of those square subimages of X whose minimum size exceeds D, where D denotes the minimum size of the subimages called also the minimum patch size. Consider then a set Ω of image descriptions of a certain type and an image descriptor F which assignes a description from Ω to each subimage from X_D :

$$F : X_D \rightarrow \Omega$$

Let a similarity relation S_F be defined via F for each pair of the subimages :

$$xS_F y \Leftrightarrow S(F(x) , F(y)) = 1$$
$$x,y \in X_D$$

where

$$S : \Omega^2 \rightarrow \{0,1\}$$

is a similarity function. The binary relation S_F is symmetric, reflexive and non-transitive.

If X_D has at least two elements and

$xS_F y$ holds for all $x,y \in X_D$,

then we call X <u>homogenous texture with respect to F and S</u>.

The minimum subimage size D can be estimated by investigating the variance of the texture feature values with respect to the change of the subimage size, i.e. by computing F for a range of sizes. The specifics of a particular texture analysis problem to be solved must also be taken into account.

2. Texture mosaic segmentation.

The problem of the segmentation of a texture mosaic may be formulated as the partitioning of the mosaic image into disjoint homogenous texture segments. The classical formulation of the problem assumes that each pixel is assigned a distinct segment label. Let X be the mosaic image and X_i topologically connected mosaic subsets, i = 1,2,...,N, where N is the number of the subsets. Introduce a <u>textural homogeneity predicate</u> H(Q;F,S) : H(Q;F,S)=TRUE means that $Q \subseteq X$ is homogenous texture w.r.t. F and S. The <u>classical statement</u> of the mosaic segmentation problem [3] is as follows:

Partition X into X so that

1. $\bigcup_i X_i = X$;
2. $\forall_{i \neq j} \; X_i \cap X_j = 0$;
3. $\forall_i \; H(X_i;F,S) = TRUE$;

4. $\forall_{i \neq j} H(X_i \cup X_j; F, S) = \text{FALSE}$

if X_i and X_j are adjacent in X.

Since under these conditions N is not unique, additional task-dependent conditions, such as minimizing N, may be also imposed.

The minimum patch constraint leads to an ambiguity in the partition which is inherent in texture segmentation. Feature values computed in the vicinity of the mosaic borders form no cluster in the feature space. The borders should be treated as discontinuity regions with the uncertain value of the features. These regions are usually a small (but not negligible) part of the image. The pixels which belong to the ambiguity zones are not assigned any distinct segment label. The new statement of the mosaic segmentation problem that takes into account this intrinsic ambiguity is as follows:

Partition X into X_i and B (boundary region) so that

1. $(\bigcup_i X_i) \cup B = X$;
2. $\forall_{i \neq j} X_i \cap X_j = 0$;
3. $\forall_i X_i \cap B = 0$;
4. $\forall_i H(X_i; F, S) = \text{TRUE}$;
5. $\forall_{Y \subseteq B} H(Y; F, S) = \text{FALSE}$;
6. $\forall_{i \neq j} H(X_i \cup X_j; F, S) = \text{FALSE}$
   if X_i and X_j are adjacent in X;
7. $\forall_i H(X_i \cup B; F, S) = \text{FALSE}$.

Condition 5 means that no subset of the ambiguity zone is texture. The task-dependent additional condition is usually to minimize the size of the boundary region.

Recognizing the ambiguity as a logical consequence of the minimum patch constraint and explicitely including the boundary region B into the segmentation algorithm leads to an improvement in the segmentation results (see, for example, [4]).

3. Texture imperfections.

In automated surface inspection, one often faces tasks that combine texture analysis with image segmentation but differ from the traditional segmentation by texture discussed in the previous section. It is necessary to detect local imperfections (defects) in uniform texture fields. The size of the defects is usually so small that, at a normal resolution, they cannot be treated as distinct texture patches. They are just "bad" regions surrounded by "good" texture. In other words, they are non-textured patches which locally break the homogeneity of a texture pattern.

The concept of texture imperfection has been originally discussed in [1]. Consider an image X which is not texture w.r.t. a feature set F and a similarity measure S. If, after exclusion of small size non-textured regions, the rest of the image becomes a homogenous texture w.r.t. F and S, the excluded regions are called local texture imperfections in X with respect to F and S. Formally, the problem of detection of a single texture imperfection is stated as follows:

Partition X into I (imperfection) and $X \setminus I$ so that

1. $I \subset X$ is a connected region;
2. H(I;F,S) = FALSE ;
3. H(X;F,S) = FALSE ;
4. $H(X \setminus I;F,S)$ = TRUE ;
5. size of I = minimum ;

Condition 2 does not yield condition 3 since the dimension of I may be less than D. Obviously, the imperfection size should be minimized.

Imperfections resemble the borders of a texture mosaic in the sense that they are also ambiguity zones of texture features. However, they differ from the borders in their size and surroundings. While an ideal border may have a close to zero width and has different textures on its sides, a texture imperfection has a finite size and is surrounded by a uniform texture.

Texture defects may have a blob-like or linear shape. They may originate from a local geometrical distortions of a uniform pattern, or be the result of a local process involving considerable intensity modifications. The minimum patch size D of the surrounding texture plays the key role in the basic classification of the types of texture imperfections (see [1] for the details). Designing an operator for defect detection in textures one must also take into account the magnitude of D.

Conclusion.

Texture description (F), similarity relation (S_F) and the minimum patch are the basic components of the operational definition of homogenous texture. Since the operational approach assumes no "absolute" description, the same image may be viewed as a homogenous texture or not depending on the choice of F and S.

The minimum patch constraint exhibits itself in the inevitable ambiguity inherent in segmentation by texture. This ambiguity must not be neglected. It should be taken into account explicitely in order to improve the segmentation results. Estimation of the minimum patch size is also necessary for the proper design of an operator for defect detection in textures.

Acknowledgement.

The author thanks J.Foglein for fruitful discussions.

References

[1] Chetverikov,D. (1985). Texture imperfections. Presented at the BPRA 3rd Int.Conf., St.Andrews, Scotland. (to appear in Pattern Recognition Letters, 1987).

[2] Gagalowicz,A. and S.de Ma. (1982). Synthesis of natural textures. Proc. 6th Int.Conf.Pattern Recognition, Munich, Germany, pp.1081-1086.

[3] Horowitz,S.L. and T.Pavlidis. (1974). Picture segmentation by a directed split-and-merge procedure. Proc. 2nd Int.Joint Conf.Pattern Recognition, pp.424-433.

[4] Tucker,J. (1985). New Shape and Texture Features From Interval Coded Images. Presented at the BPRA 3rd Int.Conf., St.Andrews, Scotland.

TOWARD A TESTBED FOR EVALUATION OF EARLY VISUAL PROCESSES

Heiko Neumann, H. Siegfried Stiehl[1)]

1. INTRODUCTION

A substantial number of international research publications on e.g. processes for low- and intermediate-level computational vision may be characterized as follows according to their style

a) quite comprehensible (though now and then incomplete) mathematical notations for the continuous case;
b) missing details on discrete implementations, e.g. numerical accuracy and stability, implicit assumptions;
c) only rather phenomenological understanding of pictorial results is mostly possible by readers;
d) only minor guarantee of own practical verification of implementation and results due
   - to lacking details of parametrization of methods
   - to concealed assumptions, limitations, restrictions, deficiencies, and weaknesses of methodologies and algorithms
   - to relatively arbitrary choice of test images for performance evaluation;
e) precise comparative/competitive evaluation of different methods for the same problem is scarcely possible;
f) missing proof of reproducible and reliable application to (in the ideal case) all potentially possible members of an image class;
g) no clear judgement with respect to the "general-purpose computational vision vs. domain-specific engineering feat" dichotomy.

Recently, other authors [5],[13] pinpointed similar critical issues of computational vision research e.g. with regard to the practical scrutiny of implementations and their illustrative results. Consequently, the necessity arises for an experimental testbed which allows for a gradual increase of the complexity of scientific problems associated with early visual information processing. Explicit control of physical phenomena in a scene, which give rise to image-domain primitives, has to be a key element in such testbeds. Moreover, only an appropriate design of experiments, e.g. by taking into account knowledge about the underlying degrees of freedom of the various physical phenomena in scenes, will guarantee results aimed at both meaningful evaluation and reliable verification/ falsification of methods.

[1)] Technische Universität Berlin, Fachbereich Informatik, FR 3-3
Franklinstr. 28-29, D-1000 Berlin 10

We describe an experimental testbed which has been particularly designed to ensure reproducable quantification and control of scene, object, and sensor parameters. Scene parameters refer to spatial arrangement of 3-d objects, local/global contrast, presence of cast shadows, as well as to locus/orientation of artificial light sources with nearly homogeneous radiance and quasi-parallel white light emission. Object parameters imply local/global geometry of solids and reflectance of non-textured surfaces. Sensor parameters are primarily specified by the characteristics of the monochromatic CCD camera and photogrammetrically determined calibration constants. Control of surface texture, disparity, and trajectories may be added according to demand.

Camera images of laboratory scenes as well as synthetic images (e.g. either to model contrast, discontinuities, and noise or to artificially generate images with Computer Graphics methods) can be produced to study both isolated physical phenomena and the interplay of phenomena in a scene. Various early vision processes, e.g. for edge detection based upon partial derivatives of regularization operators, and for global/local shading analysis have been (and will be) implemented on a VAX 750.

With regard to the development as well as to the comparative and competitive evaluation of computational vision processes, the proposed testbed promises to serve as a first step toward an eventually standardizable research environment.

2. HIERARCHY OF TEST DATA

For the purpose of evaluation of early visual processes we have defined 4 classes of test images which substantially differ in their complexity

1) synthetically generated digital images to model different features and properties to occur in digital images (e.g. idealized edge profile types, line junction types, contrast, noise etc.);
2) artificial digital images (e.g. of parametrized 3-d geometric primitives such as elliptical, cylindrical, hyperbolical, sinusoidal surfaces) synthesized with Computer Graphics methodologies for 3-d geometric modelling, orthographic/perspective projection, and surface rendering [3];
3) artificial digital images generated by a synthetic image generation testbed [4] which allows for generation of images from synthetic scenes (e.g. exhibiting arbitrary configurations of different objects, each with different material properties and arbitrary surface texture, illuminated by one or several artificial light sources);
4) "real" CCD camera images of laboratory scenes of restricted complexity.

Class 1 images allow for evaluation of processes as applied to isolated image-domain features/properties (e.g. edge profiles, junction types) under idealized as well as noisy conditions. Class 2 images are amenable to study both i) the dependency of local discrete image function properties on idealized 3-d surface geometry and illumination; and ii) the performance of processes as applied to selected isolated image features/properties under again idealized as well as noisy conditions. Class 3 images ensure evaluation of processes which, for instance, infer (prior synthesized) scene-domain properties from corresponding , e.g. idealized or noisy, image-domain properties. In addition, class 3 images render possible both to model appropriate combinations of physical scene phenomena and to investigate interactions between the resulting image-domain features. Notwithstanding the inherent restrictions of synthetic scenes and their Computer Graphics rendered images (class 2 and 3), as compared to the real world these image classes allow for at least partial analysis of the behavior and performance of computational vision processes. Therefore, class 4 images of laboratory scenes open up investigations into the performance of processes given distorted digital image domain properties which directly correspond to prior known physical phenomena in a reproducable laboratory world. Obviously, class 4 images bridge the existing complexity gap between class 3 images and camera images of real world outdoor scenes, the latter which are currently mainly used for experimental evaluation.

The proposed categorization provides a useful hierarchy to study performance of visual processes under gradually increasing complexity of test images.

3. COMPONENTS OF THE EXPERIMENTAL LABORATORY TESTBED

The testbed has been particularly designed to generate class 4 images of laboratory scenes. In this chapter we stress the choice of a light source, the calibration of a CCD camera, and the determination of surface reflectance properties.

The selection of an artificial light source has been made according to two basic requirements

a) utmost homogeneous distribution of illumination, and
b) quasiparallel emission of white light

within a certain range, e.g. of 30 cm diameter, of the illuminated laboratory world. These requirements ensure keeping of photometric assumptions necessary for global/local shape-from-shading processes [6],[12]. Moreover, violations of such assumptions may be simulated and thus the impact of such violations on 3-d shape recovery can be investigated.

Two light sources have been tested with respect to the two criteria given above: A slide projector (24 V with a 250 W halogen bulb and 90 mm/150 mm lenses) and a Kobold light source (200 W Kobold bulb). In an experiment, samples of illumination distribution have been measured within unit squares defined by equidistant grid points on a plane parallel to the light source. The distance between the slide projector light source and the plane was 2.14m. The measurements x_i for both the single central square and the central 3 x 3 squares have been evaluated for homogeneity using a common measure of distribution

$$d = (x_{max} - x_{min}) / x_{mean}.$$

Due to the measured insufficient homogeneity of the Kobold light source, only the two lenses of the slide projector have been experimentally examined with regard to their ray divergence angle (utilizing again the unit squares described above). Results showed superiority of the 150mm lens positioned at a 7.60m distance from the illuminated scene (e.g. with a divergence angle of 1.1° for the central unit square; see Table 1).

For generation of class 4 images, a SONY XC37 CCD camera with a 16mm lens is currently used (512x491 pixel resolution with 8 bit intensity resolution per pixel). The objective determination of the internal and external orientation of the CCD camera [14] is a necessary prerequisite for the definition of an accurate camera model. Such a camera model provides indispensible parameters required for e.g. recovery of invariants from linear or structural perspective [2]. Camera calibration has been carried out with the aid of the Direct Linear Transformation (DLT, see [1]). A specially designed calibration frame with attached markers of known spatial coordinates (embedded in a reference world coordinate system) has been imaged from three different fixed camera viewpoints two times each. Utilizing the DLT along with the explicitely known marker coordinates and their interactively determined image plane coordinates, the internal and external camera orientation parameters can be precisely computed (see Table 2).

To quantify surface reflectance properties of solid objects in the laboratory world, the luminance distribution of a variety of diffuse and glossy materials have been measured by means of a goniophotometer. In particular, the indicatrices of planar material samples have been measured given three selected light incidence angles ε (see Table 3 for material sample "matt silken-sheen finish lacquer"). In the main, two criteria have been taken into consideration to assess the measurements

a) optimal separation of diffuse and specular reflectance component, and
b) ideal hemispheric emission of incident light according to the theory of Lambertian reflection.

For all the currently considered 6 material samples, an approximate Lambertian reflectance has been experimentally corroborated for small light incidence angles $\varepsilon \leq 45°$. Details of experiments and results will be published in [11].

4. CONCLUSIONS AND PERSPECTIVES

In Chapter 1 we motivated the necessity for the design and realization of an experimental testbed useful for evaluation of early visual processes. In Chapter 2 we sketched the required image classes and associated experimental tasks. The components of the testbed have been briefly described in Chapter 3. Based upon both the expounded arguments and the resulting testbed, at present the following processes are being evaluated (with respect to e.g. validity in the case of noisy discrete image functions, numerical accuracy and stability, performance under noisy "real world" conditions, behavior under violations of assumptions, etc.)

a) edge detection/localization based upon first [8] and second derivatives [10],[7],
b) grouping of edge pixels into contour segments and line junction types e.g. based upon perceptual organization [9],
c) recovery of invariants (such as vanishing points and the horizon line) from linear perspective [2], and
d) local shading analysis [12] and global photometric stereo [15].

With regard to the development as well as to the comparative and competitive evaluation of computational vision processes, the proposed testbed promises to serve as a first step toward an eventually standardizable research environment.

ACKNOWLEDGEMENTS

Major parts of this research have been supported by a DFG grant for SFB 203 at TU Berlin. One of the authors (H.N.) is indebted to Prof. J.Albertz (Inst. für Geodäsie und Photogrammetrie), Prof K.Stolzenberg and Dr. G.Geutler (Inst. für Lichttechnik) at TU Berlin for discussions and help. A. Koschan kindly helped manage the text editor layout.

LITERATURE

[1] Abdel-Aziz, Y.I., H.M. Karara: Photogrammetric Potentials of Non-Metric Cameras. University of Illinois at Urbana-Champaign, Department of Civil Engineering: Report UILU-ENG-74-2002 (1974)

[2] Barnard, S.T.: Interpreting Perspective Images. SRI International AI Center: Technical Note 271 (1982)

[3] Blinn, J.F.: Models of Light Reflection for Computer Synthesized Pictures. Proceedings ACM SIGGRAPH '77 (1977), 192-198

[4] Hall, R.A., D.P. Greenberg: A Testbed for Realistic Image Synthesis. IEEE Computer Graphics and Applications 3 (1983), 8-20

[5] Haralick, R.M.: Computer Vision Theory: The Lack Thereof. Proc. 3rd Workshop on Computer Vision - Representation and Control, Bellaire/USA (1985), 113-121

[6] Horn, B.K.P.: Understanding Image Intensities. Artificial Intelligence 8 (1977), 201-231

[7] Jackèl, D., A. Koschan, H. Neumann: Design of Operators for Image Processing Based on the Bézier Approximation. Proceedings 2nd International Conference on Automatic Image Processing CAIP '87, Wismar/GDR (1987, this issue)

[8] Korn, A., C. Erdtel: Kombination verschiedener Filterkanäle zur Optimierung einer Merkmalsrepräsentation im Bildbereich. Informatik-Fachberichte 107. Springer-Verlag, Berlin (West) 1985,107-111

[9] Lowe, D.G.: Perceptual Organization and Visual Recognition. Stanford University, Department of Computer Science: Report STAN-CS-84-1020 (1984; published by Kluwer 1985)

[10] Marr, D., E. Hildreth: Theory of Edge Detection. Proceedings of the Royal Society of London 207 (1980), Series B, 187-217

[11] Neumann, H.: Evaluierung früher visueller informationsverarbeitender Prozesse in einer Experimentalumgebung (Arbeitstitel). TU

Berlin, FB Informatik: Dissertation (1987, in Vorbereitung)
[12] Pentland, A.P.: Local Shading Analysis. SRI International AI Center: Technical Note 272 (1982)
[13] Price, K.E.: I've Seen Your Demo; So What? Proc. 3rd Workshop on Computer Vision - Representation and Control, Bellaire/USA (1985) 122-124
[14] Rüger, W., J. Pietschner, K. Regensburger: Photogrammetrie. VEB Verlag für Bauwesen, Berlin (DDR) 1973
[15] Woodham, R.J.: Analyzing Curved Surfaces Using Reflectance Map Techniques. in: P.H. Winston, D. Brown (Eds) Artificial Intelligence - An MIT Perspective. MIT Press, Cambridge/USA 1979

	Lenses	
	150mm	90mm
Homogeneity of illumination $d = (x_{max}-x_{min})/x_{mean}$		
(1) central grid square	0.082	0.078
(2) central 3 x 3 grid square	0.191	0.376
Divergence angle of radiance α_i (for 30cm scene illuminated field) [°]		
(1) central grid square, α_1	1.1	2.1
(2) central 3 x 3 grid square, α_3	3.3	6.4
Distance between light source and objects for illumination of a scene range of 30cm diameter [m]		
(1) central grid square, d_1	7.60	3.90
(2) central 3 x 3 grid square, d_3	2.35	1.35

	150mm lens		90mm lens	
	 α_1^{150}	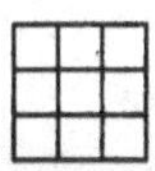 α_3^{150}	 α_1^{90}	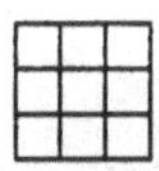 α_3^{90}
Circular unit surface patch, 1cm diameter [°]				
${}^{\bullet}\tau_1^j = 2\cdot\alpha_1^j/30$	0.0733	0.2200	0.1400	0.4267
Square unit surface patch, 1cm diameter [°]				
${}^{\blacksquare}\tau_1^j = 2\cdot\sqrt{2}\cdot\alpha_1^j/30$	0.1037	0.3111	0.1980	0.6034

Tab.1: Statistics of the properties of the slide projector light source, (a) homogeneity of illumination, radiance divergence, and distance of light source for experimental setups (b) radiance divergence measured on illuminated surface patches of unit area

Sensor calibration parameters	Position 1		Position 2		Position 3	
	Image 1	Image 2	Image 1	Image 2	Image 1	Image 2
External orientation parameters						
1. Rotations (x-,z-,y-axis) [gon] = 10/9 [°]						
ω (Omega)	-1.57	-1.58	-1.62	-1.61	-1.61	-1.63
κ (Kappa)	-0.73	-0.73	-0.76	-0.75	-0.55	-0.54
φ (Phi)	0.08	0.07	0.08	0.06	-0.29	-0.25
2. Coordinates of center of projection [m]						
X_0	11.18	11.20	11.02	11.05	10.88	10.90
Y_0	12.63	12.65	12.88	12.89	13.11	13.12
Z_0	9.09	9.09	9.18	9.18	9.29	9.29
Internal orientation parameters						
1. Image coord. of piercing point [pixel]						
x_0	245	237	255	241	259	236
y_0	262	248	266	275	232	247
2. Camera constant (lens focal length: 16mm) [mm]						
c	16.87	16.79	16.21	16.35	16.14	16.15

Tab.2: Result of camera calibration to determine external and internal orientation parameters using the method of DLT [1]

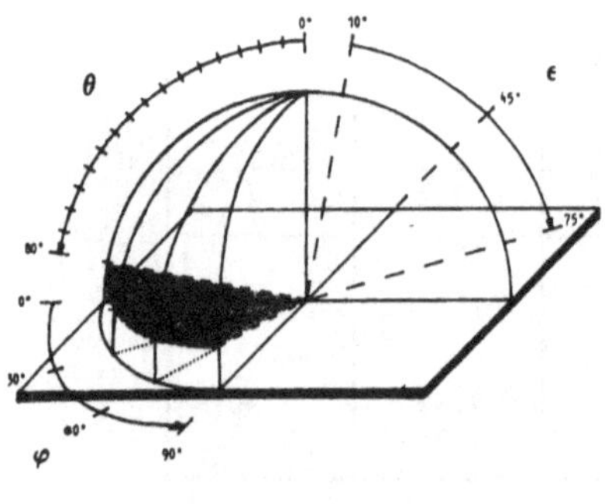

ϵ : light source incident angle (Epsilon) (measured with respect to surface normal)

(φ, θ) : polar coordinates (Phi, Theta) to define sensor position (due to assumed isotropic scattering behavior, measurements are taken over only one quarter of the hemisphere)

Paramters of the sensor:

distance from material sample d = 0.81 m
view angle α = 0.1° (= 6′)

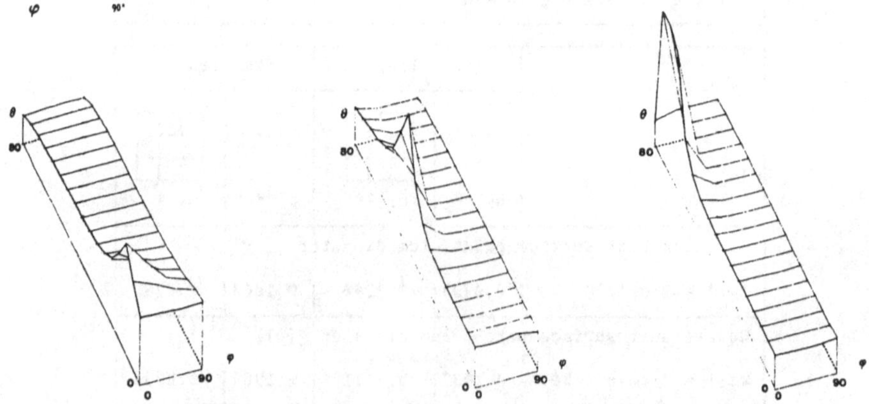

Tab.3: (a) Geometry for measurement of luminance distributions for different material samples
(b) Plot of luminance distribution (in cd/m²) for material sample "matt silken-sheen finish lacquer" for (i) ε = 10° , (ii) ε = 45° , (iii) ε = 75°

DISCRETE INTEGRAL GEOMETRY

K.VOSS [1)]

1. Homogeneous neighborhood structures

In another paper of this volume, "The three basic formulas of neighborhood structures", it was shown that there are only three homogeneous structures (P,N,C) relevant to image processing: hexagonal structures with $\nu = 3$ neighbors of each point and a mesh length $\lambda = 6$, quadrilateral structures with $(\nu,\lambda) = (4,4)$, and triagonal structures with $(\nu,\lambda) = (6,3)$.

There exist topological mappings f on these structures defined as follows: Each point $p \in P$ is mapped onto a point $q = f(p)$ such that the relation $q' = f(p') \in N(q)$ holds for the neighborhood N(q) if $p' \in N(p)$.

The concatenation of two topological mappings is also a topological mapping. Each topological mapping is either an elementary mapping (translation or rotation) or can be build up from elementary mappings by finite concatenations.

Any finite subset F of the point set P of a neighborhood structure (P,N,C) is called a figure. Each topological mapping of a figure from one place of the neighborhood structure to another (with or without rotation) is called a motion of the figure.

Let F be any figure of the structure (P,N,C) characterized by values (ν,λ), and let F' be any other figure movable by topological mappings. There are cases with $F \cap F' = \emptyset$, and there are cases with a nonempty intersection $F \cap F' \neq \emptyset$. Each figure, i.e. F or F' or $F \cap F'$, has some topological features:

- ε the number of points of the figure,
- $\varkappa$ the number of edges of the figure,
- f the number of core meshes ("inner" meshes),
- r the number of border meshes ("outer" meshes),
- k the number of components of the figure.

Let $\alpha(F)$ be any topological feature of figure F. Then we want to determine the expression

$$S_\alpha = \sum_{F \cap F' \neq \emptyset} \alpha(F \cap F')$$

1) Friedrich-Schiller-Universität, Sektion Technologie, E.Thälmann-Ring 32, Jena 6900, DDR

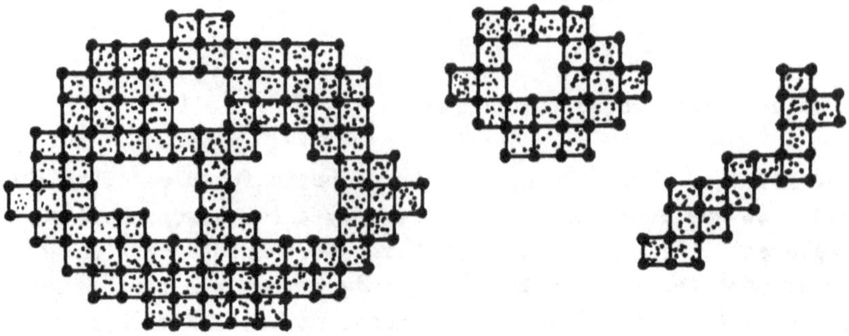

Fig.1 - These three figures have the topological feature tupels $(\varepsilon, \varkappa, f, r) = (127,213,84,4), (36,56,20,2), (29,42,14,1)$, and it is always $k = 1$.

with F as a figure fixed in the neighborhood structure, and F' as a movable figure. The sum is taken over all different positions (locations and orientations) of F' for which the intersection $F \cap F'$ is a nonempty set.

2. Basic equations of discrete integral geometry

Let us assume that α is the measure for calculating the number of points, and let F' be a figure consisting only of a single point with $\varepsilon' = \varepsilon(F') = 1$, $\varkappa' = f' = 0$, $k' = 1$. This figure F' can be placed in ν different orientations as any figure. Therefore it holds that $S_\varepsilon = \nu\varepsilon$ with ε as the number of points of figure F. Using this result, we can measure the "size" of $F \cap F'$ for any figure F' by counting the hits of a point with the intersection $F \cap F'$.

In the sum S_ε, we can take into account first the hits between the movable test point $\{p\}$ and the fixed figure F. The intersection $F \cap \{p\}$ is either a one-point set $\{p\}$ or an empty set. Now we count the hits between this pointlike intersection and the movable figure F'. From that, the first equation

$$S_\varepsilon = \sum_{F \cap F' \neq \emptyset} \varepsilon(F \cap F') = \nu\varepsilon\varepsilon'$$

follows. Because of the analogy of this equation with the known expression $S_a = 2\pi aa'$ for the similar area problem of real integral geometry (i.e. continuous integral geometry), we call the described method "discrete" or "discontinuous" integral geometry (see L.A.SANTALO, Integral Geometry and Geometric Probability, Addison-Wesley Publ.Comp. 1976).

Now let α be the measure for calculating the number of edges, and let F' be a figure consisting only of a single edge with $\varepsilon' = 2$, $\varkappa' = 1$, $f' = 0$, $k' = 1$. The edge can be arranged in two different orientations. Each orientation and each location of F' which gives an edge as intersection $F\cap F'$ counts as one in the sum $S_\varkappa$. Therefore $S_\varkappa = 2\varkappa$ with $\varkappa$ as number of edges of figure F. Using this result, we can measure the "edge size" of $F\cap F'$.

In the sum $S_\varkappa$, we take into account first the fixed figure F. The intersection $F\cap\{p,q\}$ with the edge $\{p,q\}$ is either an edge or a single point or an empty set. Only the edges are of interest. Now we count the hits between these intersection edges and the edges of the movable figure F'. We obtain the second equation

$$S_\varkappa = \sum_{F\cap F'\neq\emptyset} \varkappa(F\cap F') = 2\varkappa\varkappa'.$$

Finally let α be the number of core meshes, and assume that F' is given by a single core mesh, i.e. F' is a figure which consists of λ points and λ edges. The figure F' can have λ different orientations at the same place. In the way described for deriving the first and the second equation, we obtain our third equation

$$S_f = \sum_{F\cap F'\neq\emptyset} f(F\cap F') = \lambda f f'$$

The similar structure of these three equations follows from the fact that the "topological size" of a figure F can be measured by any of the three features ε, $\varkappa$, or f.

3. Examples

These three equations are the basic equations of discrete integral geometry because other measures or features of interest can be determined by ε, $\varkappa$, and f. For example, the equation $\varepsilon - \varkappa + f + r = 2k$ holds for all planar figures constructed from k components with a total of r border meshes.

The figures F and F' be squares of $\varepsilon = \varepsilon' = n^2$ points in the orthogonal grid. What is the sum S_ε of all points of all intersections $F\cap F'$? From the first equation it follows that $S_\varepsilon = 4\varepsilon\varepsilon' = 4n^4$. These squares contain $\varkappa = \varkappa' = 2n(n-1)$ edges and $f = f' = (n-1)^2$ core meshes. For all components of all intersections we have $r = k$. Therefore $k = 2k - r = \varepsilon - \varkappa + f$ holds, and finally we obtain

$$S_k = \sum_{F \cap F' \neq \phi} k(F \cap F') = S_\varepsilon - S_\varkappa + S_f$$

The number of all components of all intersections is now determined by

$$S_k = 4n^4 - 8n^2(n-1)^2 + 4(n-1)^4 = 4(2n-1)^2 ,$$

and the mean size of a component is $\bar{\varepsilon} = S_\varepsilon / S_k = n^4/(2n-1)^2$.

This result can be generalized. If F and F' are planar connected figures with $k = k' = r = r' = 1$, then all intersections are also connected figures with a single border mesh. Therefore in general we obtain

$$S_k = \nu\varepsilon\varepsilon' - 2\varkappa\varkappa' + \lambda f f'$$

for any two figures F and F' and any of the three neighborhood structures characterized by (ν, λ). The mean "topological sizes" of the intersection components are

$$\bar{\varepsilon} = \nu\varepsilon\varepsilon'/S_k \quad , \quad \bar{\varkappa} = 2\varkappa\varkappa'/S_k \quad , \quad \bar{f} = \lambda f f'/S_k \ .$$

It will be the aim of investigations in the next future to derive further equations of discrete integral geometry, and to solve practical problems of stochastic grid geometry. The "normal" continuous integral geometry may be a valuable heuristic tool for this work.

VISION FOR INDUSTRY

K.Fritzsch 1)

1. Introduction

Machine vision represents a pool of methods and devices for estimating the state of industrial environments and objects . At the present time, we are in a transition stage where the "machine-vision-for-industry"-activities in the R&D field of the last years are on the way to give real economical effects for users and manufacturers of vision systems indicating the establishment of machine vision in industry [8],[9],[10],[19],[20],[34],[35],[39],[42],[45].

Usually, it is necessary that the research results related to hardware as well as to software are reworked and completed for broad industrial application. In the competition between different sensory principles vision as a non-contact and microcomputer based method is well appropriate for flexible automation. Optimistic expectations at the early eighties have not been fully confirmed, but now there are reliable forecasts that in the nineties worldwide accumulated sales of industrial vision systems will reach figures over 10 billions US S.

In GDR, R&D for machine vision is well established. The engagement of manufacturing engineers and economists must be intensified to achieve significant benefits by large scale application of machine vision in industry. About 3 conferences with machine vision as a main topic are held each year. The same figure is found for conferences where machine vision is included. Mainly two journals serve the publication needs of the field. The annual rate is 8 to 10 papers in each journal. Some general literature is available[5],[6],[17],[26].

2. System architecture

In accordance with the stages of processing, a visual system consists of a number of modules, each for solving a certain processing subtask: irradiation, image acquisition and formation, storage, preprocessing, recognition, and interpretation.

This modularity will remain relatively stable in the future development, at least in its functional aspect, although the single modules will be

1) Academy of Sciences of GDR, Central Institute for Cybernetics and Information Processing, Kurstrasse 33, DDR-1086 Berlin

further improved to give higher performance, higher reliability, and smaller size on the base of advanced physical realization. The overall system design needs not to be modular if a compact implementation is more desirable. A short description of the modules is given in the following:

- Irradiation: Spatial relations between source of radiation and camera (back, front, lateral irradiation) have significant influence on the image quality; structured irradiation may help in obtaining spatial information and in discriminating an object from the background. Polarized irradiation gives good results in the case of specularly reflecting objects but the shading of the backgruond by the object and the degradation of intensity of irradiation by the filters are often severe disadvantages. The design of the scene is of equal importance. A well chosen presentation of the objects in the visual field can save much processing power.
- Image acquisition and formation: VLSI- sensor chips as CCD-lines and CCD- matrices are dominant because of high geometric accuracy, low-voltage power supply, miniaturization, external controllability, price degression etc.

Raster grids of 256x256 to 1024x1024 pixels with digital radiometric resolution from 1 to 6 bit and scanning rates between 3 and 5 Mcps are mostly used. Line sensors are bounded in the scanning rate because of the necessary high intensities of irradiation. The trend is towards shading-corrected digital cameras with quadratic raster grid. Miniturization has led to a camera with a weight of 115g and a size of 35x45x75 cm .

The silicon substrate of the sensor is responsible for the high sensitivity in the near infrared, which has advantages (well matched to incandescent light) and disadvantages (decrease of geometrical resolution with longer wavelength) too.

- Image storage: Usually, the data rates for the image acquisition and for the image processing are very different. In such cases, an image frame buffer is needed. For binary systems, buffering may be limited to several pixels (to be coded into one memory word) or to a raster scan line. High performance systems (if not run in pipeline) are equipped with sufficently large frame storage capacity.
- Preprocessing: The modules of these stages are of central importance, realizing local operators for restoration, edge detection etc. and performing the feature extraction. Preprocessors may be implemented as dedicated hard ware, e.g for the very efficient run-length coding of binary images or for the local operations in windows of small size (3x3 to 11x11).

In high performance systems programmable preprocessors (with dedicated software tools) are mostly used guaranteeing a high degree of flexibility [27],[37],[49].

- High-level processing: In this last stage image data is already very compressed, but the algorithms have no longer such a kind of simplicity and repeatability as in low-level processing. Therefore universal processors possibly on the base of bit-slice components are a convenient solution.

From the pure hardware standpoint there are three main modules: camera, frame store with preprocessor, and high-level processor with process interface which are connected by a video channel or video bus and a system bus resp.

In the last 12 months, add-in boards became broadly available for widely used busses and personal computers as well giving the possibility to configure systems depending on the needs of application[43].

3.General functional demands

Vision systems are designed for acquiring morphometrical, radiometrical and geometrical quantities from which under industrial conditions process information may be derived, e.g. length, area, surface property, object position, shape features, simultaneously assuring a high degree of reliability and reproducability. The distinguishing property as compared to common techniques of measurement and supervision is the ability to extract objects of interest and relating morphometric quantities. Patterns and shapes prove to be relatively stable against variations of irradiance, luminance and position especially in the two-dimensional case. The pattern recognition ability may be used also for estimating the object-dependent coordinates in case of dimensional measurements. Therefore, pattern recognition is a basic technique in machine vision. Optimum performance parameters vary with application: positional and orientational accuracy of +- 1% and +- 1 degree resp. in robot vision, +- 1 % in length measurement, discriminating power for 20 classes of objects on the average (maximum 1000 !), overall processing times between 10ms and 3s.

A major lack of the present day's machine vision systems is their limited ability to process 3D-information. Using structured irradiation is succesful only in special cases.

4. High-performance systems

High-performance systems find their way to broad use, especially for problems with high geometrical (up to 1000x1000 pixels) and radiometrical (up to 256 levels) resolution, for large portions of imaging, for motion and color analysis[25]. If high savings in production costs can be achieved, then expensive visual inspection is accepted. High-performance systems were initially developed for analysing remote sensing data of terrestrial resources and objects. In a parallel processor from GoodYear

Aerospace 6550 MOPS fixed point or 216 MOPS floating point are performed. A japanese system contains frame storage of 1.6 MByte. By reason of their special architecture the image processing performance of these and other systems exceeds the capabilities of the extremely expensive supercomputers.

Inspection of photomasks for integrated circuits, of thick film circuits, and of printed circuit boards are typical applications[1],[32],[33]. High-speed implementation of simple single-pixel or neighbourhood algorithms, specialized feature extraction and image analysis with extensive software support and communication to CAD data bases as well are the main features for such vision systems, which are often tailored by the supplier to the special application.

5. Visual sensors

In the late seventies a worldwide development of medium-performance machine vision systems has begun, utilizing the price and size degression of microelectronic components. The aim was to build general purpose systems for demonstrating typical solutions for part recognition, template matching, and scene analysis on the base of binary images[14],[15],[28]. An important feature of these systems with a microprocessor as the main processing resource is using the run-length code for binary images in all cases of relatively smooth object contours and no pepper-and-salt noise.

Evidently, demonstration with the aid of costly computers alone proves not be sufficient. The industrial user, on the other side, was more interested in the role such systems could play in solving sensory problems of real automation projects, including flaw detection, gaging, and identification. In a trial-and-error process a considerable degree of correspondence between the user's demands and the vendor's offers has been achieved now. Users are prepared to provide for special installations, e.g. irradiation, object presentation etc. Vendors do much to define the environments the system can be tailored for, to made calibration tools available, and to give extensive software support for better communication.

Because of the rather different complexities of the tasks the availability of low-, medium- and high-performance systems is necessary for economical reasons. Worldwide, about 100 to 150 systems are commercially available.

The autonomous behaviour of vision systems is achieved by foregoing calibration and teaching. Well-developed methodologies are at hand for two-dimensional binary images. For gray-level processing and 3D-recognition much operator interaction is necessary with present systems. Model- and knowledge based approaches seem to be more adequate.

6. Evaluation criteria

From the user's standpoint the following criteria for evaluating vision systems may be applied:

- amount of additional installations as irradiation, fixtures and so on,
- facilities for connecting the system to different cameras,
- controllability of image acquisition,
- 2D-, 2.5D- or 3D-processing,
- real-time behaviour,
- resolution and accuracy,
- reproduceability of processing results,
- facilities for calibration,
- facilities for maintenance and trouble- shooting,
- tolerable environmental conditions,
- references to realized applications.

Vision system are suited for fixed automation(inspection, acquisition of state variables) and for flexible automation (recognition, derivation of control quantities) as well. For the latter case

- modularity and configurability

may be desirable properties in addition.

An important feature of the systems is their software support[46],[48]. The wide variety of software approaches may be explained by the relatively embryonic state of the field and by differing requirements of application. Almost each supplier offers its own software package. Ratings for programming the systems range from Very Easy(if menu-driven) over Not So Easy (if a set of predefined commands is available) to Complex (if the user would have to program the sytems by a high-level instruction set). It is desirable to include modules for the standard preprocessing functions and to help the user in developing long series of processing steps.

7. Typical applications

Among the diversity of potential applications there are some very prospectous ones which became operational at several places exhibiting the expected benefits, e.g for measurement, template matching, and part recognition. At the same time, some pretententious solutions are prepared for industrial use[2],[3],[7].

In the following some typical applications for the first-generation binary systems are listed:

- Template matching[40],[41]:

Estimation of position and orientation of silicon chips before wirebonding by search for and matching of partial patterns.

The identity of type-writer types is checked before assembly using special irradiation and template matching inside of windows defined during teaching. High-volume products as fasteners, cups, and closures are inspected before further processing or assembly. With the aid of a multi-camera system the accuracy of video monitors is measured.

- Recognition of parts for industrial robots[18],[30],[31],[50]:

Classification of parts hanging on a conveyor is performed to call the corresponding program of a painting robot. The trade-off between misrecognition and rejection is solved in advantage of the latter.

In the automotive industry, wheel assembly is guided by vision to recognize the orientation of the wheel hub and to tell it to the robot.

- Dimensional measurement combined with shape analysis:

First applications were realized in fixed coordinate systems[16].The object-dependent coordinate system defined during recognition may be used as the frame for performing length measurements on the object. Non-contact gaging of complex shapes as gears or doors becomes possible.

- Dynamical measurement:

An impressing example using a line camera is estimating the vibrations of type writer wheel spokes with a time resolution of 26 ms and an accuracy of 10 um[11].

Grey-level processing is recommended when, in the above mentioned cases, the preconditions for binary image acquisition cannot be guaranteed. But often it is mandatory as the in following cases:

- Dimensional measurement in the sub-pixel range:

An important step was the success in developing equipment for geometrical measuring with sub-pixel accuracy. By means of analyzing the grey-level distribution across an edge it is possible to achieve an resolution of about 10 to 20 times better than it would be on the base of pixel counting [23], [24],[51].

- Spatial measurement:

High-performance systems are used for photogrammetrical estimation[21] of spatial coordinates of bodies, blades, buildings etc. in the machine and ship building industry. The main problem is the recognition of corresponding points in the image from different standpoints of cameras. Accuracies of 0.1 mm are achieved. Matching with or derivation of a CAD description may be included.

Analysis of interferograms is another possibility, especially for final inspection. Non-destructive testing of tyres with the aid of holographic interferograms is done using the system A6472 from Kombinat robotron[12]

- 3D-recognition:

One-camera systems can extract spatial information if the scene is irradiated with structured light in form of stripes or grids. For example, in arc-welding the (horizontal) deviation of the seam and the (vertical) offset of the workpiece is determined with the aid of stripe irradiation. Accuracies below 0.1 mm are possible.

- Printed circuit board inspection:
Component identifcation, solder-joint and copper-trace inspection, lead-through-the-hole verification are typical tasks. They are rather different from the image-processing point-of-view. Therefore several methodologies must be used at the same time. But the peculiar difficulties stem from the fact that there are very large variations among the class of defectless products. To avoid to much false alarms it is necessary in many cases to classify and not only to locate each defect. This field has been studied very intensively and not been solved satisfactorily yet.

8. Present and future developments

Recent progress in acceptance of machine vision is encouraging for the vision community. Large-scale transition to grey-level and colour processing are the issues for next time. Processing performance for 3D-scenes will be improved. The limited ability to process 3D-scenes is a main drawback for broader utilizaion in robotics.The extensive activities necessary for programming, calibration and teaching to each new installation will be reduced by the step-by-step introduction of model- and knowledge-based processing.

Including nonvisual knowledge sources may give better performance as it was successfully demonstrated in recognizing asbestos fibres[4]. As already indicated in discussing printed circuit board inspection one weakness of present days systems is their limited ability to avoid false alarms. Using information from X-ray analysis the amount of false alarms could be diminished significantly in recognizing.

Broad use of robot vision means that 15-20% of all robots are equipped with it. Similar figures have to be achieved in the other fields of automation. With the retrofit approach mostly used at present this purpose may be reached only partially. In robotics some integrated systems are operational now. On this base operator communication, teaching, and maintenance become easier improving the economic benefits from this emerging technology.

In the following a number of examples is listed where vision systems became operational in GDR:

- recognition of parts hanging on a conveyor to be feeded to a painting robot
- sorting of cutting blades for reaping-machines by aid of special marks
- scaling of pipettes
- measuring the thickness of wires and cables (at some places in the electrotechnical industry)
- localisation of hot glass pipes to be attached to thermos vessels.

Some factories are preparing equipment for automatic recognition and contactless measurement which will become operational in the next months.

Especially recommended is planning applications with CCD-line cameras because of cost effectivenes and availability (in form of the system BES2ØØØ with the BEE1Ø1Ø of VEB Studiotechnik).

Thanks are due to U. Schroeder for technical assistance in preparing the paper.

References:

[1] Adaway,W.G.L.: Automatic visual inspection applied to silicon chip packaging. SPIE 376(1983),Optical Sensing, 1Ø3 - 1Ø7

[2] Aleksander,I. et al.: Computer vision systems for industry: WISARD and the like. In: Real-time image processor architectures for use in computer vision and robotics. Computing and Control Division, London 1982, 4/1 - 4/13

[3] Ansorg, J. u.a.: Einsatz einer CCD-Matrixkamera in der Justage optischer Bauelemente. Bild und Ton 38(1985)12,37Ø - 374

[4] van Antwerpen,G. et al.: Automatic recognition of asbestos and other mineral fibres. Proc. 8th Int. Conf. Pattern Recognition, Paris Oct. 1986, 555-557

[5] Autorenkollektiv: CCD-Bildsensortechnik. Technische Hochschule Ilmenau/ KdT Bezirksvortand Suhl 1985

[6] Autorenkollektiv: KdT-Leitfaden zur industriellen Bilderkennung. Kammer der Technik Berlin 1987

[7] Baker, L.R.: Optical scanning in product inspection. SPIE 378(1983) Optical Sensing, 58 - 64

[8] Batchelor, B.G. et al.(eds.): Automated visual inspection. IFS-(Publ.), London 1985

[9] Brady,M. and R.Paul: Robotics Research. The First Intern.Symposium. MIT Press, Cambridge(Mass.) 1984

[1Ø] Conference Proceedings of the Third Annual Applied Machine Vision. Schaumburg(Ill.)1984

[11] Dressler, L. and S. Hecht: Schwingungsmessung mit CCD-Zeilenkamera. Feingeraetetechnik 35(1986)12, 538-54Ø

[12] Eichhorn, N. et al.: Digital analysis of tyre interferograms. Acta Polytechn. Scandinavica, Appl. Phys.149, Image Science '85, 88 - 91

[13] Emmett,A.: Stereodigitized measurements of auto parts. Comp. Graphics World. Oct. 1985,33 - 36

[14] Fimmel, B. u.a.: Ein Baugruppensystem zur mikrorechnergesteuerten Bildaufnahme und - verarbeitung fuer die Automatisierungstechnik. Bild und Ton 38(1985)12, 361 -363,381

[15] Franke, K.-H.: Das visuelle Erkennungssystem BS2 - ein Schritt zur Schaffung komplexer Automatiserungsloesungen. 29. IWK ILmenau 1984, Bd. 3, 195 - 198

[16] Frielinghaus, K.-O. und S. Hecht: Laengenmessung mit CCD-Linienkameras. Bild und Ton 37(1984)7, 2Ø5 - 2Ø8

[17] Fritzsch, K.: Visuelle Sensoren - Algorithmen und Systeme. Akademie-Verlag, Berlin 1987

[18] Fritzsch, K.: Visuelle Systeme fuer Industrueroboter der 2. und 3. Generation. Beitraege zur Optik und Quantenelektronik 8(1983)14-17

[19] Fu, K.S.: Pattern recognition for automatic visual inspection. Computer (Dec. '82), 34 - 4Ø

[2Ø] Gomersal, A.: Machine intelligence: an international bibliography. IFS (Publ.)& Springer-Verlag,Berlin 1984

[21] Haggren, H.: Real-time photogrammetry. Acta Polytechn Scand. Appl. Physics, Image Science '85, 5 - 8

[22] Hantke, D. u.a.: CCD-Zeilen sind Praezisionsmassstaebe.. Feingeraetetechnik 34(1985)7, 29Ø-292

[23] Hegedus;Z.S. and G.W.Small: Shape measurement in industry with subpixel definition. Acta Polytechn. Scand.Appl.Phys.,Image Science '85
[24] Hollinghum, J.: Vision sensing provides automated calibration. Sensor Review(July 1983)137
[25] Ito,T.: Digital color picture processing. In: Kant,M. and F. de Coulon: Signal Processing. Theory and Applications. North Holland, Amsterdam, 198Ø, 71 -81
[26] Kammer der Technik: Kurzreferate der 1. Internat. Fachtagung Automatische Bildverarbeitung. Berlin Okt. 1985
[27] Kempe, V. et al.: The interactive image processing console A6471. Proc. 6th ICPR, Munich 1982,6Ø7-6Ø9
[28] Kriesten, St. u.a.: Geraetetechnik zur Bildverarbeitung mit dem Mikrorechner k152Ø. Bild und Ton 36(1983)12,37Ø-373
[29] LaCoe, D.: In-process vision inspection of automotive tie-rods. Sensor-Review (April 1984)74 - 77
[3Ø] Leichsenring, O. and Chr. Peterson: Robot vision paves the way for flexible workshop automation. ASEA Journal 57(1984)8-13
[31] Morris, H.M.: Robot vision: bringing eyesight to the blind. Control Engng.(Jan. '85), 64 - 7Ø
[32] Odawara, G. et al.: An integrated pattern inspection system. 8th Int. Conf. Pattern Recognition Paris 1986, 5-8
[33] Okamoto, K. et al.: An automatic visual inspection system for LSI photomasks. Proc. 7th ICPR Montreal 1984, 1361 - 1364
[34] Proceedings 15th ISIR, Tokyo 1985
[35] Pugh,A.(ed.): Proc. 4th RoViSeC, London Oct. 1984, North Holland, Amsterdam 1984
[36] Radig, B.: 2D- und 3D-Objektbeschreibung fuer Sichtsysteme. VDE-Fachberichte 35, Mustererkennung 1983, 5.DAGM-Symposium, VDE-Verlag, Berlin 1983, 3ØØ - 318
[37] Rebel, B. and W. Wilhelmi: Trends in image processing architecture and implementation. In Herrmann,R.(ed.): Automatic image analysis in medicine and biology - results and progress. Leipzig 1986
[38] Roesler, U. u.a.: Digitale Bildverarbeitung und -auswertung fuer die Kontrolle von Schneidwerkzeugen. Bild und Ton 38(1985), 375 - 378
[39] Rosenfeld, A.: Machine vision for industry: concepts and techniques. Robotics Today Dec. 1985, 19-22
[4Ø] Schoenborn,W. u.a.: Teile-und Lageerkennung mit dem System MAVIS. Bild und Ton 38(1985)12, 357 - 36Ø
[41] Schwarz, W.: Visuelle Qualitaetspruefung und - sicherung nach dem Binaerbildverfahren. Bild und Ton 88(1985)12,364 - 366
[42] Tech Tran Corporation: Machine vision systems: a summary and forecast. Tech Tran Corp., Napperville(Ill.) 1983
[43] Titus, J.: Versatile add-in boards acqyire and process video-images and graphics in real-time. EDN(Aug. 1Ø86)1Ø3-11Ø
[44] Uebel, W.: Visuelle Inspektion metallisch blanker zylindrischer Oberflaechen. Bild und Ton(to be published in 1987)
[45] Vision '85, Conference Proceedings. March 1985, Detroit(Mich.). North Holland, Amsterdam 1985
[46] Voss. K. et al.: Interactive software sytems for computer vision. In: Progress in Pattern Recognition 2. North Holland Amsterdam 1985
[47] Waldman, H.: Update on research in vision. Robotics Today (June 1983), 63 - 67
[48] Westendorf, Chr.-M.: Bildverarbeitungssoftware auf der Basis von FORTH. Bild und Ton (to be published)
[49] Wilhelmi, W.: Ein System zur digitalen Echtzeitverarbeitung von Bildern. Beitraege zur Optik und Quantenelektronik 8(1983)181-184
[5Ø] Wloka, K.: Visuelle Qualitaetsbeurteilung von Rohrboegen mittels geometrischer Objektvermessung. Bild und Ton 38(1985)12,367 - 369
[51] Woschni, H.-G. et al.: Praezisionslaengenmessung mit Hilfe von CCD-Zeilen. Wiss. Beitraege IH Wismar(1984) Sonderheft 4, 61-66
[52] Wurmus, H.: Anwendung von CCD-Zeilenkameras in der Automatisierungs- und Robotertechnik. radio-fernsehen-elektronik 33(1984)7,451-453

IMAGE PROCESSING FOR THE VEGA-MISSION TO COMET HALLEY

M.Danz, G.Elter, T.Mangoldt, D.Möhlmann, B.Rubbert and U.Weidlich [1)]

1. Introduction

Two of the VEGA-2 spacecraft close encounter images of comet P/Halley are most appropriated for studying properties of the cometary nucleus. The images were taken on the 9th of March 1986, 1.5 seconds before ("image 1") and 98.7 seconds after closest encounter ("image 2") from distances of 8,045 km and 11,060 km, respectively. The angle spacecraft-nucleus-Sun for these two images was 28.4° and 23.0°, so that both images show the dayside of the cometary nucleus with its great activity. Consequently, the nucleus is visible only through the scattered light from the near-nucleus dust, which is most intense above the "active" dayside. Special image processing software has to be used to identify shape of and structures on the cometary nucleus. This processing and the resulting physical properties are described in the following paper.

2. Image processing

2.1. Restauration

There are 4 lines of reduced sensitivity of the CCD-Matrix in each of the both images. For our first investigations they have been "filled" by averages of neighbouring real lines. A further step could be to "compensate" the reduced sensitivity by shifting the intensity (brightness) in these lines up to the level of the neighbouring lines of the image.

2.2. Noise reduction

There also exists some noise on the images. In order to get images for physical interpretation the noise has to be separated. This has been performed by using an algorithm, described by Richter (Ref.1), which is based on a variable two-dimensional filter. This filter fits automatically its own band width and is based on a modified Haar-transformation. The optimal separation of the noise is realized then by a further variable digital filter (two-dimensional Kalman filter), having

[1)] Institut für Kosmosforschung der AdW der DDR, Rudower Chaussee 5, Berlin 1199, DDR

a band width which is steered by the above mentioned Haar filter.

2.3. Maximum gradient

To detect structures the local maximum gradient is computed for every pixel from the original data and "noise-reduced" then by the above described algorithm. From the so processed images we got clear indications for structures on (or directly above) the surface, for a strong decrease of brightness connected with the real curvature of the nucleus toward the limb (and terminator) - which appears as a bright seam of greatest gradients in brightness on the processed images -, and for single jets, directed into dayside hemisphere. But it is difficult to derive the real contour of the nucleus from the "gradient-images". This is due to the fact, that for relatively low optical thickness (corresponding approximately to a brightness $\sim\cos^2\alpha$ with the zenith-angle α) the gradient is greatest inside the real limb.

2.4. Laplace transformation

Additional information on shape of and structures on the nucleus can be found by the help of Laplace-transformed images, showing the values of the second derivatives. Figure 1 represents the Laplace-transformed "image 1". White values correspond to relatively great positive curvature (in the brightness field), while black represents the relatively great values of negative curvature. The outer white "seam" around the

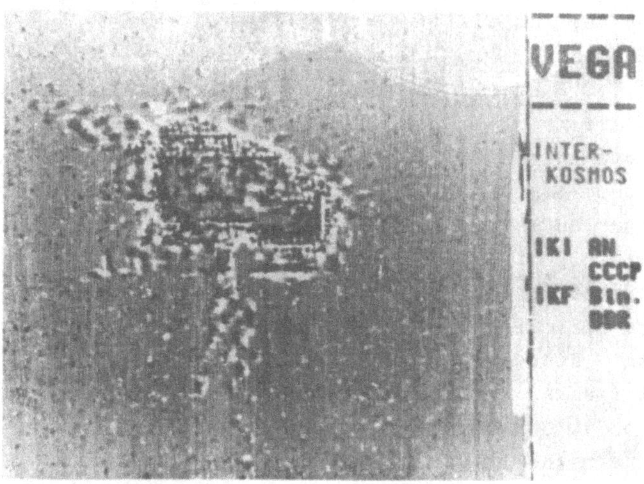

Fig.1 - Laplace-transformed "image 1".

nucleus should indicate the position of the limb. Additionally, structures can be seen at the surface of the nucleus, which differ in their morphology from the eventually not yet removed noise outside the nucleus. Similarities are possible with structures, seen in the regions of jets. So these structures should be related to those at the active surface or to structures, connected with outgassing and jets above the surface.

2.5. Overlay

A synthesis of the above mentioned "noise-reduced", gradient, and Laplace-transformed images is possible, using the three independent processor pipelines (red, green, blue) of a special display processor for simultaneous visualization of up to three different images. On such an overlay one can see, for example, the linear and ring-like structures of the nucleus region, resulting from the Laplace-transformed image, in blue colours and the bright seam of steepest brightness gradients in red colours together with the "noise-reduced" original image in green colours. The resulting image is useful in order to get a full impression of the cometary nucleus with the features on the surface and the near-nucleus jets.

3. Physical properties

The processed images show shape and indications for surface structures of P/Halley. Using a truncated cone capped at both ends by hemispheres to approximate the nucleus the derived dimensions are 15.7±0.6 km for the overall long axis, 7.5±0.3 km for the diameter of the larger hemisphere, and 5.0±0.2 km for the diameter of the smaller hemisphere. The uncertainties are mainly due to the fact that the contour of the cometary nucleus can not be identified directly, but only via maximum gradient and Laplace transformation. The most appropriate method to determine the contour seems to be the Laplace transformation, indicating different curvatures in the brightness field. A strong positive curvature can be expected near the limb, where the brightness should decrease (in outward directions) as it is indicated by the maximum gradient. At the limb this decrease in brightness should change strongly due to the different (weaker) radial dependence of the scattered light outside the cometary body. Consequently, a strong positive curvature can be expected at or near the limb.

The above given values are determined in this way.
Furthermore, the Laplace-transformed images show indications for large-scale features on the surface of P/Halley. Figures 2 and 3 give a gra-

phical representation of these features, identified on the two VEGA-2 images taken 1.5 seconds before and 98.7 seconds after closest encounter. On the "second image" the structures are shifted with respect to the "first image" due to the motion of the spacecraft. The good agreement between the expected and the real shift of the structures may serve as an indication for their reality.

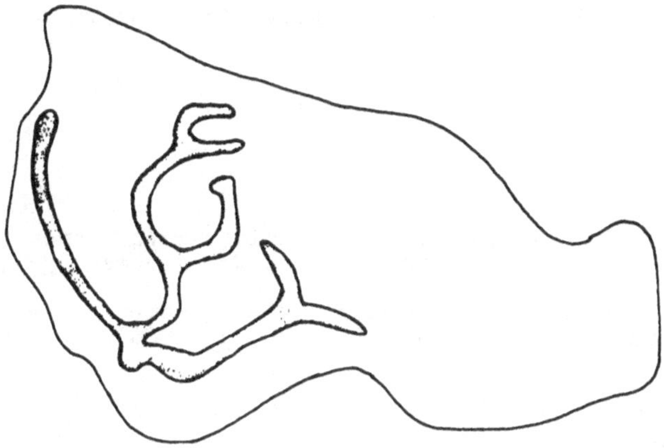

Image 1

Fig. 2 - Surface structures seen on the VEGA-2 image which was taken 1.5 sec. before encounter.

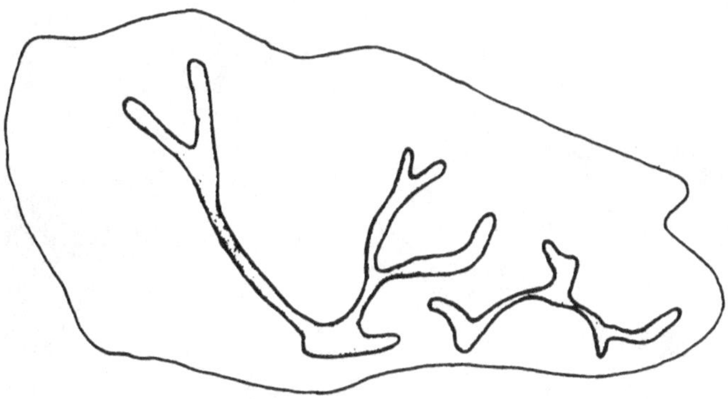

Image 2

Fig. 3 - Surface structures seen on the VEGA-2 image which was taken 98.7 sec. after encounter. They are shifted with respect to Fig. 2 due to the motion of the spacecraft.

A further indication for the reality of these structures stems from their similarity to the distribution of active lines (footpoints of jets), derived by Sekanina and Larson (Ref.2) from jet observations of P/Halley in 1910. From this fact one can also conclude that these line-type areas of enhanced activity (outgassing) are of more permanent character. Consequently, they should be connected with structures on and beneath the cometary surface and may serve as a first indication for an internal structure of the cometary nucleus, which, for example, could consist of a small number (of the order of 10) of large km-sized "building blocks". The boundaries of these blocks could be the line-type areas of enhanced activity.

4. References

/1/ Richter, G.M.: Zur Auswertung astronomischer Aufnahmen mit dem automatischen Flächenphotometer. Astron. Nachrichten 299 (1978)6, 283-303.

/2/ Sekanina, Z. and Larson, S.M.:Dust jets in comet Halley observed by GIOTTO and from the ground. Nature 321 (1986)6067, 357-361.

Acknowledgement

The authors wish to thank Drs. H.Lorenz and G.M.Richter, Zentralinstitut für Astrophysik der AdW der DDR, R.-Luxemburg-Str. 17a, Potsdam-Babelsberg 1502, DDR, for their extensive support to the image processing.

IMAGE MODELLING AND ANALYSIS IN HOLOGRAPHIC AND SPECKLE PATTERN INTERFEROMETRY

Roland HÖFLING [+]
Wolfgang OSTEN [++]

1. Introduction

Holographic interferoemtry (HI) as well as speckle metrology are well known techniques in experimental mechanics for the measurement of deformations and vibrations [5,8]. Generally, the results are light intensity distributions described by some periodical function, i.e. fringes. Those fringes are subject to digital image processing. In this paper, we give some information on our recent work in this field, using the image processing system ROBOTRON A6472. We produced both synthetic holographic interferograms and speckle correlograms modelling the physical process of image formation. Image analysing routines reduce the desired displacement information from the fringe patterns and synthetic fringes served as a convenient tool in checking analysing algorithms and in error estimations .

2. Holographic interferometry

2.1. Modelling

Two states of the coherently illuminated object under investigation are recorded on a photographic plate successively adding a reference wave in order to write a hologram . The simultaneously reconstructed light fields of both states give rise to an interference pattern, described by the intensity I(x,y)

$$I(x,y) = I_0(x,y)\, n(x,y)\, (1 + V(x,y) \cos(\delta(x,y))) \qquad (1)$$

where I_0 is is the local intensity average of the simply illuminated and observed object , n is the so called "speckle noise" considerably degrading the image , V is a week space dependent fringe visibility function caused by fringe localization effects and finally, δ is the phase we are looking for. The latter contains information on object surface displacements. In the following, we neglect the space dependence of V. All other contributions to Equ. (1) have been simulated and are shown in Fig.1.

[+] Academy of Sciences of the GDR, The Institute of Mechanics, POB 408, Karl-Marx-Stadt, 9010, GDR

[++] Academy of Sciences of the GDR, Central Institute of Cybernetics and Information Processes, POB 1298, Berlin, 1086, GDR

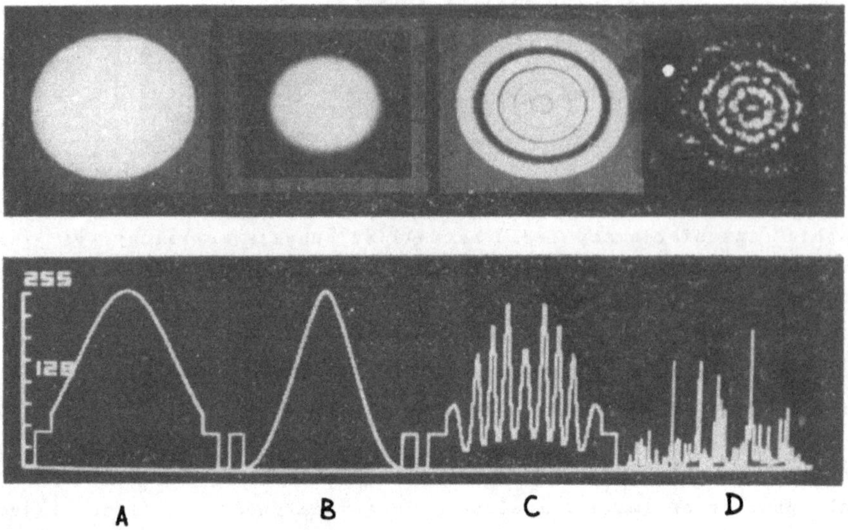

Figure 1: Formation of synthetic holographic interferograms
A: object brightness $I_o(x,y)$
B: phase $\delta(x,y)$
C: smooth fringes (without noise) $I_o \cdot (1-\cos(\delta))$
D: same as (C) but muliplied by speckle noise n(x,y)

Whereas the calculation of Fig.1(A) and (B) is straighforeward, the simulation of a speckle pattern seems worth speaking of.

In order to solve the problem of speckle pattern creation, we follow the natural process and use a simplified model of it. Speckle(strongly speaking subjective speckle) are the result of coherent superposition in the diffraction spot of an optical system. In this small area, there interfere all waves scattered in a neighbourhood of the imaged "point" and having a random phase distribution. Thats, why they may be described by addition of independent complex amplitudes yielding a circular Gaussian distributed sum due to the central limit theorem [3]. In our approach, we overlay Gaussian distributed random amplitudes A(m,n) at pixel locations (m,n) within the diffraction spot of the imaging system.

$$I(k,l) = \left| \sum_{m,n} h(k-m,l-n)\ A(k-m,l-n) \right|^2 \tag{2}$$

Approximating the point spread function h(i,j) by a radial rect-function and using the minicomputer SM4-10 it takes us about 30 min to calculate a 256 x 256 pixel pattern as shown in Fig.1(C) [4]. The mean speckle size (correlation lenth) varies with the diameter of h(x,y) and the monitor prints of three patterns with different speckle size are shown in Fig.2

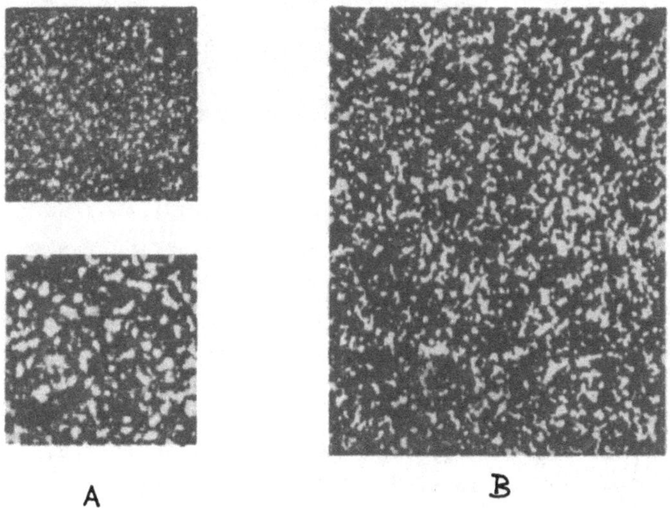

Fig. 2 : Synthetic patterns of different size (left) and real pattern (right)

A: two synthetic patterns with different speckle size

B: real speckle pattern obtained from a coherently illuminated plane object (enlarged)

In any case, the patterns satisfy a negative exponential intensity distribution. For comparision, Fig. 2 shows a real speckle field, produced by diffusely scattered laser light.

2.2 Analysis

Equ.(1) reveals the troublesome fact that the phase $\delta(x,y)$ cannot be taken from the measured value I(x,y) at point (x,y) only, because of the unknown functions I_o , n, V . The speckle noise n(x,y) may be suppressed by mean filters or a special geometric filter, discussed in section 2 and the remaining unknown quantities, beside the phase, are I_o and V . One of the solutions is the measurement of a number of intensity fields with a certain additionally introduced phase shift and eliminating I and V by combining the measurements (phase - sampling). However, this approach prerequisits something like real-time interferometry and is not

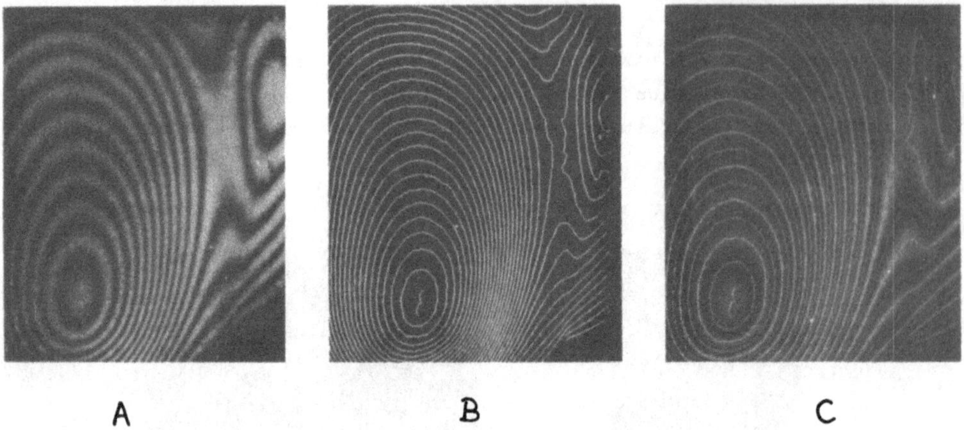

Figure 3: Evaluation example in holographic interferometry

A: monitor print of a holographic interferogram (placed at our disposel by V. Grosser)

B: automatically extracted skeleton lines (after about 35 s)

C: interferogram of (A) with overlay of improved skeleton

always applicable.We, instaed, tried to solve for the phase by searching intensity extrema which correspond to $\delta = N \lambda/2$ (N=1,2,3) if I_0 and V vary slowly.The latter is true for V, in general, but I_0 should be eliminated by a division. In order to construct I_0 , a low-pass filter has been applied that causes the fringes to blur or a convex hull may be built [6]. Sometimes, the illuminated object itself may be the source of

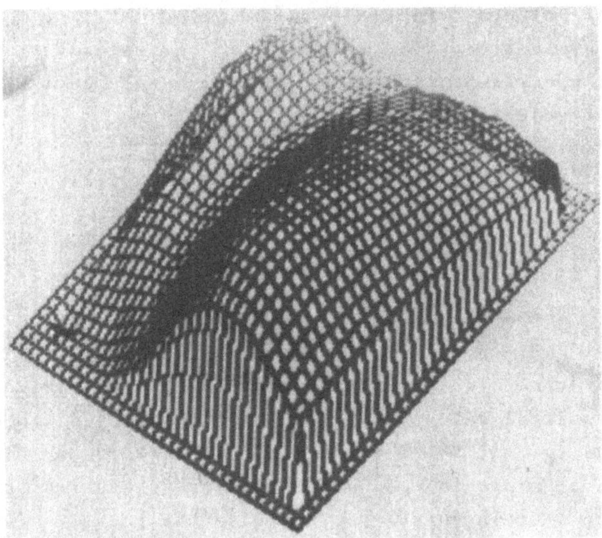

Figure 4: Plotting of the phase field obtained from the skeleton in Fig. 3(C)

⟨I⟩ . Futher processing involves an adaptive binarization [2] and line thinning extracting a fringe skeleton after less than 10s by virtue of the fast image processor the system A6472 is configured with. This automatically produced lines have to get a number (interference order) by use of atracking routine and therefore,some possible shortcomings (gaps, branchings) may be removed interactively by a comfortable skeleton improvement routine (Fig.3(B)) [7]. Finally, the phase field is interpolated on a mesh of 32 by 32 nodal points and a monitor print of its plotting is shown in Fig.4.

3. Speckle pattern interferometry

3.1. Modelling

Speckles are an image degrading noise in HI but speckle metrology, in contrast, takes advantage from the displacement information carried by that speckle pattern [5,8]. Information decoding is achieved by the crosscorrelation function of the speckle patterns corresponding to the undeformed and deformed object states, respectively [8,4]:

$$\varsigma_{12}(x,y) = \frac{\langle \Delta I_1 \quad \Delta I_2 \rangle}{\langle I_1 \rangle \; \langle I_2 \rangle} = 1/2 \; (\; 1 + \cos(\; \delta \; (x,y))) \tag{3}$$

where δ depends on the displacement field. Our interferometer allows two statistically independent speckle fields to interfere producing a resulting third pattern which interferometric properties are described by Equ.(3). For an example, we use a modified Michelson - interferometer, substituting the optical flat surfaces (mirrors) by rough ones [1]. The speckle patterns of both object states have been calculated simulating the phase change caused by a central loaded circular plate and are shown in Fig. 5(A) and (B).

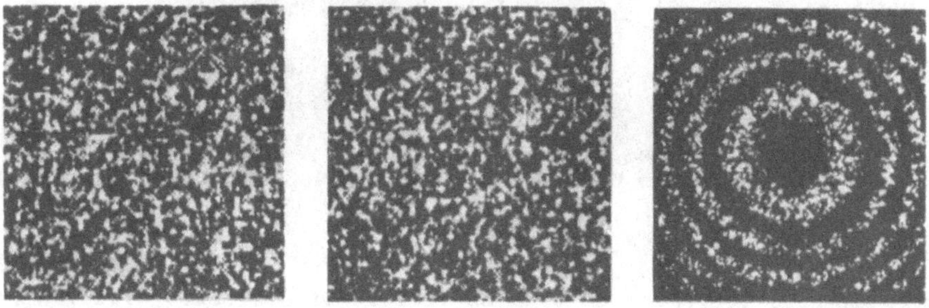

Figure 5 : Synthetic speckle pattern correlogram
A: pattern of the undeformed object
B: pattern of the deformed object
C: correlogram of the patterns in A and B : $(A-B)^2$

The crosscorrelation function is obtained by calculating the squared difference of both patterns

$$\langle (I_1 - I_2)^2 \rangle = 2 \langle I \rangle^2 \sin^2(\delta/2) \tag{4}$$

and the correlation fringes according to Equ.(4) (the correlogram) are printed in Fig. 5(C).

3.2. Analysis

In order to evaluate such correlograms as in Fig. 5(C) automatically, the high contrast multiplicative speckle noise must be removed. A common method is the application of mean filters to correlograms or phase reconstructions. Such mean filters are simple and fast but they will fail if the fringes become too dense. For illustration, the filtered images of the correlogram in Fig.5(C) are shown in Fig. 6.

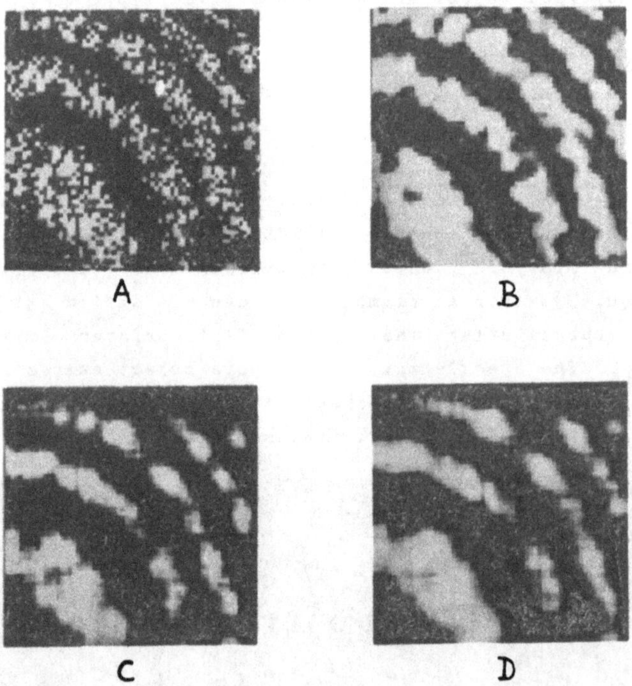

Figure 6: Speckle reduction by mean filters of different window size
A: original image (enlarged part of Fig. 5(C))
B: 3x3 window (0.08s)
C: 5x5 window (0.16s)
D: 7x7 window (0.24s)

The speckles are not removed completely even by the 7x7 window but the fringes tend to fuse already. The reslult of multiple repetitions of the mean filter is similar. For this reason, special speckle filters have been developed by several authors. Crimmins [8] proposed a geometric filter for synthetic radar applications. This filter limits the curvature of the intansity mountains and modifies image parts of high curvature only, as speckle are. We have programmed the parralel image processor to run this geometric filter algorithm and found it well suited for smoothing specke correlograms (Fig. 7).

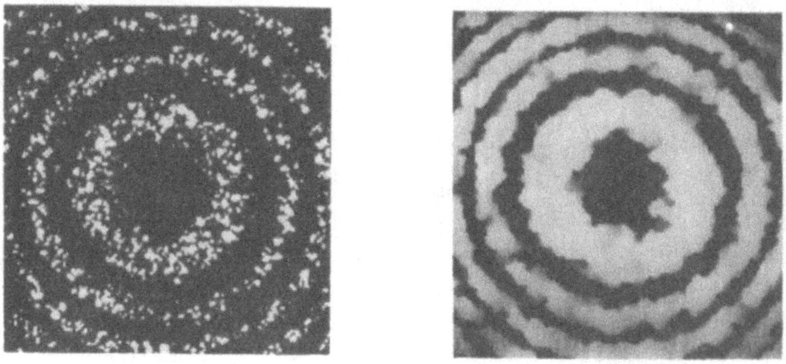

A B

Figure 7: Speckle reduction by geometric filter
A: original image
B: result of four geometric filter cycles (60s) applied to (A)

In our experimantal setup, we fed the camera signal to the A6472 image processing system where the TV frames are digitized at video rates to give a 512 by 512 pixel image, 6 bit deep. We store the reference pattern and calculate $(I_1 - I_2)^2$ with additional contrast enhancement in real time (every 40 ms). Electronic noise impairs the fringe contrast as visible in Fig. 8(A).

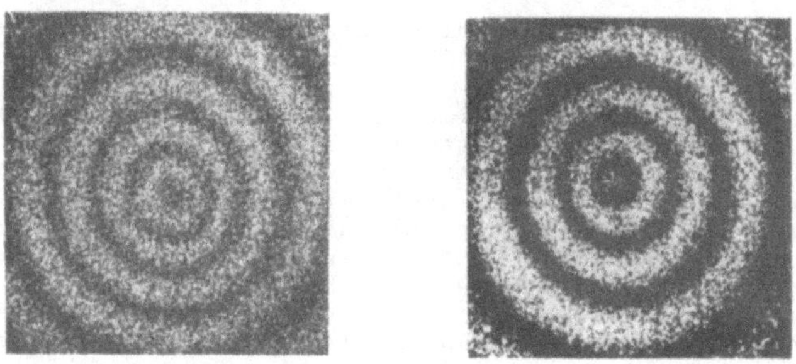

A B

Figure 8 : Electronic noise reduction by TV frame averaging
A: Correlogram obtained from singele TV frames
B: Correlogram obtained from 20 frames for each pattern

This is the reason for using the average of about 20 TV frames for at least the reference pattern. The frame averaging takes us about 2s but produces correlograms of much better contrast (Fig. 8(B)).

After speckle reduction, the intensity field compares with that of holographic interferometry (Fig. 9(A)) and the same analysis algorithm may be applied. Doing so, we have a system , capable of real-time interferometry with optical rough objects and a rapid online evaluation implemented on hardware usual in trade.

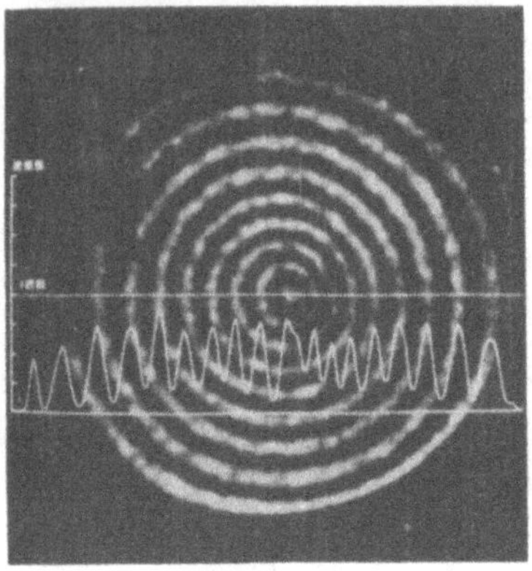

A

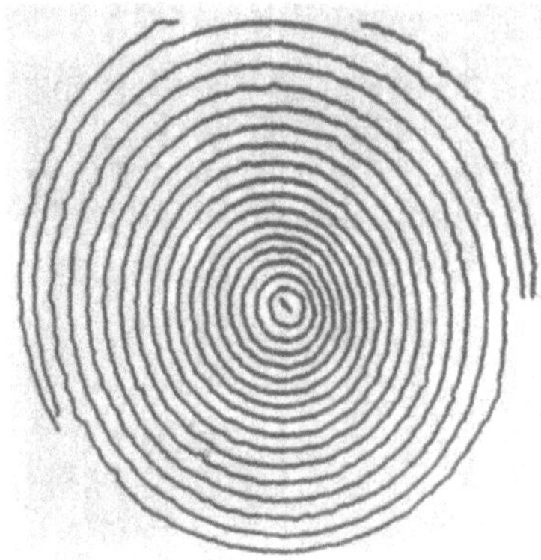

B

Figure 9: Evaluation of correlation fringes
A: correlogram after speckle reduction
B: skeleton of (A)

4. References

[1] Crimmins, T.R., Geometric filter for reducing speckle, Opt.Engng. 25(1986),651-4

[2] Eichhorn,N., Saedler,J., Wilhelmi,W., Digital analysis of tyre interferograms , Acta Polytech.Scand.Appl.Phys.Ser. PH150 (1985), 88-91

[3] Goodman,J.W., Statistical properties of Laser Speckle Patterns, in Laser Speckle and Related Phenomena (ed. J.C.Dainty),Springer, Berlin,1975

[4] Höfling,R., Osten,W., Displacement measurement by image processed speckle patterns, Optica Acta (in preparation)

[5] Jones,R., Wykes,C., Holographic and Speckle Interferometry, Cambridge University Press, Cambridge, 1983

[6] Kovalevski, W., A new computational method for shape restoration from interferograms, 1. Internationale Fachtagung "Automatische Bildverarbeitung", 17.-18.10.1986, Berlin

[7] Osten,W., Saedler,J., Wilhelmi,W., Schnelle Auswertung von Interferogrammen mit Verfahren der digitalen Bildverarbeitung, Lasermagazin (in preparation)

[8] Yamagushi,I., Fringe Formations in Deformation and Vibration Measurements Using Laser Light, Progress in Optics (ed. E. Wolf), XXII (1985), 271-339

THE MICROCOMPUTER BASED FRINGE PATTERN PROCESSING SYSTEM

Andrzej Pietrzyk[1)]

INTRODUCTION

In several methods used in experimental mechanics fringe pattern is obtained as a result of an experiment. There are displacement measurement methods based on the phenomenon of light interference, like various moire techniques, holographic interferometry or speckle. In all these methods specimen made of real structural materials are usually tested. Thus fringe pattern frequently appears on non-homogeneously bright background and frequently noise filtration has to be applied prior to further processing. Fringe patterns obtained in these methods usually constitute contour maps of some deformation characteristic, e.g. in-plane displacement component, off-plane displacement or slope of the deformed surface i.e. the derivative of the off-plane displacement. The values of the derivatives of this characteristic are to be calculated. Therefore, modelling of deformation characteristic is often used. A very popular method of (mainly qualitative) stress analysis based on the phenomenon of birefringence is photoelasticity. In this group of techniques models made of special transparent materials are often tested instead of real structures or structural members. Thus, it is possible to assure homogeneous background brightness and the initial filtration can be ommitted. Colour images are obtained and the analysis of colours can give additional information. The analysis of colour images is, however, more complicated and requires bigger memories and better video equipment. Fringe patterns obtained in these techniques constitute contour maps of the differences of principal stresses. The separation of principal stresses leads to the determination of stress state in the specimen. This is, however, a complicated task. Integration is usually required. The finite element method is often used. There are also several other methods giving visual information, although of different type (e.g. brittle lacquer technique, USG, X-raying or thermovision).

Digital computers are presently commonly used in experimental data analysis. In several applications they are also used for data logging and/or the control of testing machines. The wide dissemination of personal computers with memories ranging at 1MB and the introduction of integrated, "flash" A/D converters with sampling frequencies exceeding 20 MHz enables the automatization of visual information processing. With some kinds of computers easily available on the market it is possible to create a versatile workstation having all of the afore mentioned capabilities. The presence of a microcomputer with high processing power at the test stand

[1)] Politechnika Warszawska, Instytut Mechaniki Konstrukcji Inzynierskich, Al. Armii Ludowej 16, 00-637 Warszawa, Poland.

enables the application of hybrid experimental - numerical techniques. Minicomputers were used in laboratories for computer aided analysis of visual data. In [5] fringe patterns obtained in plane photoelastic models investigations were considered. The image was coded in the form of 512x290 pixels matrix with the resolution of 256 grey levels per pixel. Colours were not considered and noise filtration was ommitted. Fringes were recognized by a constant threshold technique. Finite element method was used for separation of principal stresses. In [11] the similar problems were considered, but the preliminary image filtration was employed. The image was coded in the form of a 1024x512 pixels matrix with 64 grey levels resolution per pixel. Colours were not analyzed. Finite element method was also used for postprocessing. In [2] the digital spatial frequencies filtering was used for smoothing the light intensity distribution in speckle images. However,* image was analyzed along lines, not as a full-field one. In [12] an interesting method for the determination of phase distribution in moire fringe patterns was proposed. It was based on the spatial frequencies spectrum analysis of the pattern brightness along the lines approximately perpendicular to the fringes. A CCD line sensor with 1024 elements providing 256 grey levels per pixel was used. However, the reports on the similar applications of microcomputers aren't yet numerous.

WORKSTATION

To make the choice of the microcomputer to be used one has to formulate the requirements it should fulfil. In our case they were formulated as follows:

- the system should be fairly cheap
- the microcomputer should be a standard, easily available one
- the system should be versatile, not restricted to image processing
- computing power should be big enough to run finite element programs
- image coding should be performed in not less then 512x512 pixels in 64 to 256 grey levels resolution per pixel
- image coding should be performed in real-time

To fulfil this requirements an IBM-PC compatible computer was chosen. Initially, the AT version was planned, but the XT was finally decided due to high cost of the image processing cards. This kind of computer is easily available and fairly cheap. It has 8 expansion slots which makes it a versatile system, easy to expand by simply building-in various available add-on cards. The software available for this computer is very rich. There are ready to use finite element programs and compilers for various high level languages. In our case the computer was provided with a 20 MB Winchester type hard disk drive, two floppy disk drives of 360Kb each, 640 kb RAM memory, dot matrix printer STAR SG-15 and an arithmetic coprocessor. The clock frequency was switchable between 4.77 and 8 MHz, what together with the arithmetic coprocessor gave high computing power, enabling effec-

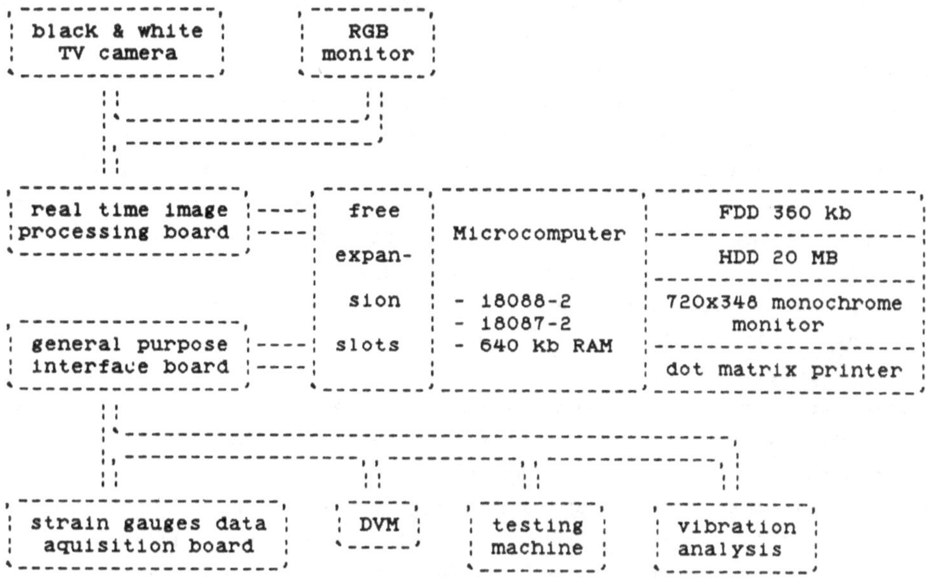

Fig. 1. Block diagramm of the system

tive utilization of finite element programs. The schematic diagram of the system is shown in fig. 1. This system was also described in [7].

To enable image processing a special add-on card fitting to a standard IBM-PC/XT or AT expansion slot was chosen - the PIP - 1024 card made by Matrox Co. This card contains a flash A/D converter, which performs digital coding of black & white TV camera signal in form of a 512x512 pixels arrays with 256 grey levels resolution per pixel in the real-time. It has also it's own memory of 1 MB, so that four subsequent TV frames can be coded one by one. Alternatively, the continuous frame grabbing mode can be selected. The board has also a set of three D/A converters that are used to display the coded image on the colour TV monitor in the so-called pseudo-colours directly when it is being coded. The data transfer between the image processing card and an IBM-PC/XT compatible computer is possible both through ports and through the DMA channel. The card contains a set of eight look-up tables for each of the three output channels and for the input channel. Look-up tables are user defined. Each of the look-up tables is 256 bytes long, so that all of the 256 grey levels available can easily and quickly be changed for displaying. Using three tables at once one can transform a black & white image into a colour one. This enables one to distinquish details with much greater ease then when the output is in grey levels only. This feature can be used to mark areas where some criterion is fulfilled, e.g. the actual plastic zones being formed during an experiment. The use of a set of three such cards together in one computer is possible. In this case one card is configured as a master and the other two as slaves. If each card is connected to one of the colour TV camera RGB channels it is possible to take a synchronized snapshot with all three cards and digitize and process colour images in real-time with the spatial resolution of 512x512 pixels and with 256 intensity levels for each of the fundamental RGB colours. Several software controlled features make the work with this card easy. One can switch the video standard between American and European, select an internal or external synchronization, select one of the three available input ports or cause the output signal to be given as input (feedback mode) and control adjustment of video input signal for optimum A/D converter range by means of two registers controlling digitally the amplification of the input analog signal and the A/D converter work. One megabyte of on-board memory is useful also in the case, when only one TV frame is grabbed. It is possible to zoom-up the digitized image and then scroll and pan it on the display. It is also possible to store four versions of the same image at once to enable easy and fast display changing. It has to be noted, that the RAM memory of the microcomputer is usually sufficient to contain only one image at a time The PIP-1024 image processing card is supplied with the software library PIP-EZ. This software can be used in one of the two modes: as an interpreter or as a routines library for high level language. In the first case, a set of commands is used to control image coding, processing, displaying and data transfer, each command being executed immediately when it is entered from the keyboard. The set of commands includes taking a snap, defining the lookup tables, controlling data transfer between the card and an IBM-PC resources. Image processing commands include histogram calculations of the defined window, histogram displaying and convolution of an image with a user defined 3x3 kernel. In case the routines library mode is selected it is possible to incorporate card control commands as subroutines into programs written in the Microsoft Corp. C or Fortran77 programming languages. Thus, it is easy to create one's own set of image processing routines. The most obvious disadvantages of the software package are: the lack of zooming routines, the lack of a cursor and it's motion on the display, the lack of user defined kernel size and the lack of the hardwired FFT routine.

IMAGE PROCESSING SOFTWARE

Foundations of the digital image processing can be found in [8]. The same author gave a brief review of the parallel algorithms for image processing in [9]. By this kind of algorithms it is possible to perform image processing in the real time, provided that special multiprocessor computers are available. These computers are, however, still very rare. Therefore, the software prepared in our laboratory, although in several steps parallel, cannot perform the task of image analysis on the personal computer in the real time. This software was discussed in [6] and therefore would only briefly be reviewed here. In general, fringe pattern analysis is divided into several steps. The first step consists in image filtration in order to suppress the noise disturbing fringe pattern. A number of alternate algorithms was prepared to perform this task (e.g. image smoothing, features detection and non-isotropic filtration, spatial frequency filtration based on Fourier analysis). The use of the particular algorithm depends on the quality of the coded image and the type of fringe pattern. It is possible to skip this step if it is not necessary. Image smoothing techniques consist in exchanging the actual pixel with the mean or mediane value of it's neighbourghs. This can be regarded as a low pass filter. Mediane filter better preserves sharp edges but it converts peaks into step-like objects thus decreasing their spatial determination. Spatial frequency filtration is effective in noise reduction for simple patterns of regular geometry i.e. for patterns with narrow spatial frequency spectrum. However, it affects fringes localization. For more complicated patterns it is difficult to fit a proper filter to avoid changing of fringe pattern itself during noise filtration. Non-isotropic filtration considered by Granlund [3], requires some knowledge on the pattern. For fringe patterns this should be performed along ranges which represent fringes in the grey level surface. Therefore, the probable directions of fringes in small windows should be known. To find them, special templetes are defined and convolved with the image. They are designed to fit ranges passing in various directions through the window. High responce means good fit of templete to the actual image. The set of directions obtained is inconsistent due to noise. To improve consistency the relaxation method [4] is used. Once a consistent set of directions is determined it is used as a mask for non-isotropic filtration. This method is particularly recommended for this kind of fringe patterns in which fringes are of the approximately the same width everywhere in the image. Unfortunately, this condition does not hold, especially for photoelastic images. The second step consists in the recognition of fringes. This is achieved either by thresholding or by maximum-searching range-tracing algorithms. The automatic threshold selection is possible by the analysis of histogram. Thus, it is possible to select different threshold value for each window accounting for the slow changes of background brightness. In case, the thresholding is used for fringe pattern recognition, the thinning algorithms transforming binary picture into eight-simple digital arcs are useful. An example of such an algorithm is described in [1]. The most important and the most difficult step consists in fringe labeling. By now, human interaction is required at this step. For some kind of pattern it seems quite a problem to distinguish peaks from valleys in the contour map obtained from experiment. The last step consists in the analysis of strains and stresses revealed by the recognized fringe pattern. Least squares surface fitting is used to obtain approximate values between the contour lines Finite element method is used for the separation of principal stresses in case of photoelastic fringe. A combination of this two methods was proposed in [10] with interesting results.

FURTHER EXTENSIONS

As it was mentioned before the currently used system is, thanks to it's free expansion slots, ready for further development. Several applications for image processing are planned concerning mainly non-destructive testing methods like X-raying, gamma-raying and thermovision. For example, the stereometric reconstruction of reinforcement bars in reinforced con crete structures investigated by gamma rays is considered. For effective analysis of photoelastic images it would be useful to expand the system by adding two PIP - 1024 cards more. This would enable the analysis of the pattern in true colours. Another development consists in the instalation of the popular IEEE-488 interface board (as it is already shown in fig. 1). This interface enables the computer to control several laboratory devices like digital voltmeters, strain gauge data aquisition systems, vibration analyzers. This interface can also be used for testing machine control during experiment. Since the system is used at the technical university, it's versatility is particularly important. It can be utilized to illustrate several novel experimental techniques in the course of education saving substantialy the university's spendings. Finally, it has to be noted that the development of hardware dedicated to image processing is rapid and new devices are being introduced on the market. They are based on the CCD matrices of 1024x1024 elements. These units are particularly suited for laboratory applications due to their high spatial resolution and to the fact that the detecting elements positions are fixed. This feature distinguish them from vidicon based cameras. Thanks to the different technology they can be expected to be cheap soon as well.

REFERENCES

[1] Arcelli C., Sanniti di Baja G.: A thinning algorithm based on prominence detection. Pattern Recognition, 13(1982)/3, 225-235.

[2] Chambless D.A., Broadway J.A.: Digital filtering of speckle photography data. Experimental Mechanics, 19(1979)/8, 286-289.

[3] Granlund G.H.: In search of a General Picture Processing Operator. Computer Graphics and Image Processing, 8(1978), 155-173

[4] Hummel R.A., Zucker S.W.: On the foundations of relaxation labeling processes. IEEE-PAMI, 5(1983)/3, 267-287.

[5] Muller R.K., Sackel L.R.: Complete automatic analysis of photoelastic fringe. Experimental Mechanics, 19(1979/7, 245-251.

[6] Pietrzyk A.: Digital image processing in experimental mechanics. Proc. VIIIth International Conference on Experimental Stress Analysis, Amsterdam, The Netherlands, May 12-16, 1986, 281-290.

[7] Pietrzyk A.: On the digital processing of fringe patterns. Proc. IMEKO 1st TC15 Symposium, Plzen, Czechoslovakia, May 25-28, 1987 (to be published)

[8] Rosenfeld A., Kak A.C.: Digital pictures processing. Academic Press,

New York, 1976.
[9] Rosenfeld A.: Parallel algorithms for image analysis. Proc. 1 Internationale Fachtagung Automatische Bildverarbeitung, Berlin, October 17-18, 1985, P8.1-P8.4.
[10] Segalman D.J., Woyak D.B., Rowlands R.E.: Smooth spline like finite element differentiation of full field experimental data over arbitrary geometry. Experimental Mechanics, 19(1979)/12, 429-437.
[11] Seguchi Y., Tomita Y., Watanabe M.: Computer-Aided FRinge-pattern ANalyzer - A case of photoelastic fringe. Experimental Mechanics, 19(1979)/10, 362-370.
[12] Takeda M., Ina H., Kobayashi S.: Fourier transform method of fringe pattern analysis for computer based topography and interferometry. J. Opt. Soc. Amer., 72(1982)/1, 156-160.

COMPUTER AIDED PROCESSING OF SEM - INTERFERENCE - PATTERNS IN FRACTURE MECHANICS AND MICROMECHANICS

Kühnert, Rolf+ Michel, Bernd+ Henniger, Karin+ and Hussack, Jürgen+

A concept of computer aided testing of construction materials is represented that permits to investigate the mechanical behaviour of fracture mechanical compact tension specimens by means of scanning electron microscopy.
It enables to find out the fields of displacement components and derivable fields and quantities (strain, stress, plastic zone dimensions, J - integral values).
The procedure will be explained in image generation and image processing as well as interpretation.

0. INTRODUCTION

A complete characterization of the mechanical quality of construction materials seems to be a necessary understanding for the optimum use of these materials today more than ever.
The application of Fracture mechanics is to develope and apply methods and techniques allowing to make statements about reliability of structural parts considering that they include imperfections like cracks or crack-similar defects.
Thereby the defect is estimated quantitativly analysing the displacement- and stress behaviour in the nearest crack tip surroundings -often by means of finite element or boundary element methods- and calculating the so-called characteristic quantity of the choosen fracture criterion on the base of these field quantities.
Then reliability statements are derived from the comparision of the obtained characteristic quantity and its experimental determined critical value.
Investigations about validity of special criterions in the field of fracture mechanical basic research as well as the determination of critical values in the field of application desires an experimental approach to the values of characteristic quantitiy at different load steps.

+ Academy of Science of the GDR, The Institute of Mechanics
PF 408 Karl - Marx - Stadt 9010

One possibility therefore is offered by means of the Micro Moiré Technique (MMT) as special realization variant of moiréinterferometry using features of a scanning electron microscope (SEM).
This Technique has been developed at The Institute of Mechanics /1/.

1. IMAGE GENERATION

Moiréinterferometry denotes the analysis of interference phenomenons hereafter the superposition of a quasiperiodic object structure associated with the object state to be analysed with a reference structure regarding the selected reference state and similar to the object structure so that the phase of resulting interference pattern contains the wanted measurement.
The term quasiperiodic structure is used for the position dependence of a physical quantity that can be written in fourier representation with position dependent coefficients and spatial frequencies.
The basic idea of the scanning electron microscopical realization of moiréinterferometric displacement measurements is that the (periodically modulated) yield of secondary electrons is only ejected at discrete (periodically arranged) geometrical points fixed by the primary probe regime what leads to the generation of a moiré interference pattern about the image of microscopic surface cut-out (i.e. the crack tip surroundings) on SEM - CRT if suitable imaging parameters have been used.
The superposition of two periodic structures

$$S_p(\underline{r}) = \sum_{n=-\infty}^{+\infty} P_n(\underline{r}) \exp\{-2\pi i n \hat{\underline{p}}(\underline{r}+\Delta\underline{r}_p)\} \qquad (1)$$

and

$$S_b(\underline{r}) = \sum_{m=-\infty}^{+\infty} B_m(\underline{r}) \exp\{-2\pi i m \hat{\underline{b}}(\underline{r}+\Delta\underline{r}_b)\} \qquad (2)$$

with $\underline{r}$ = plane vector of position

$\hat{\underline{p}} = \hat{p}\underline{e}_p = \underline{e}_p/p$ = spatial frequency of object structure

$\hat{\underline{b}} = \hat{b}\underline{e}_b = \underline{e}_b/b$ = spatial frequency of reference structure

$\Delta\underline{r}_{p,b}$ = phase shifting terms

with local extremums at $\underline{R}_p$ and $\underline{R}_b$

$$(\underline{R}_p + \Delta\underline{r}_p)\hat{\underline{p}} = G_p \quad / \quad G_p \in \{G\} \qquad (3)$$

$$(\underline{R}_b + \Delta\underline{r}_b)\hat{\underline{b}} = G_b \quad / \quad G_b \in \{G\} \qquad (4)$$

leads for multiplicative as well as additive combination to the build-up of local extremums of the structures

$$S_{\Sigma}(\underline{r}) = S_p(\underline{r}) + S_b(\underline{r})$$

$$S_{\Pi}(\underline{r}) = S_p(\underline{r}) \times S_b(\underline{r})$$

at the geometrical points $\underline{R}$ with $\{\underline{R}\} \subseteq \{\underline{R}_p\} \cap \{\underline{R}_b\}$.
If the imaging system interpolates linearly between nearest extremums $\underline{R}_i$, $\underline{R}_j$ interference fringe patterns with the spatial frequency

$$\hat{\underline{f}} = \hat{f}\underline{e}_f = \underline{e}_f/f$$

arise.
If $\underline{R}_i$, $\underline{R}_k$ are extremums at nearest neighbouring fringes the projection condition

$$(\underline{R}_i - \underline{R}_k)\,\hat{\underline{f}} = 1 \tag{5}$$

is current.
Then the vectorial statement

$$\hat{\underline{f}} = a_1\hat{\underline{p}} - a_2\hat{\underline{b}} \quad / \quad a_1, a_2 \in \{P\} \tag{6}$$

leads to

$$a_1(G_{pi}-G_{pk}) - a_2(G_{bi}-G_{bk}) = 1 .$$

On the other hand

$$1\,(G_{pi}-G_{pk}) - n_2(G_{bi}-G_{bk}) = 1 \iff \hat{p} > \hat{b}$$

$$n_1(G_{pi}-G_{pk}) - 1\,(G_{bi}-G_{bk}) = 1 \iff \hat{p} < \hat{b}$$

with $n_1, n_2 \in \{N\}$ is valied, so that the spatial frequency of interference structure can be written as

$$\hat{\underline{f}} = a_1\hat{\underline{p}} - a_2\hat{\underline{b}}$$

$$/ \; a_1 a_2 = a_1 + a_2 - 1 \; ; \; a_1, a_2 \in \{N\}$$

(a_1, a_2 = thinning-out coefficients /2/).
The constancy of fringe order $M(\underline{R}) = \underline{R}\hat{\underline{f}}$ / $M \in \{P\}$ and $\Delta M = \pm 1$ between neighbouring fringes for all (!) fringe points $\underline{R}$ can be tested using

(3), (4) and (6).
If there appears a sufficient faint position dependence of the spatial frequency of one of the superimposed structures (i.e. $\hat{\underline{p}}'(\underline{r})$) a position dependent deformed interference structure will be generated and if

$$\hat{\underline{p}}'(\underline{R})\underline{R} = \hat{\underline{p}} \cdot (\underline{R} - \underline{u}(\underline{R})) \tag{7}$$

it can be described as

$$M(\underline{R}) = \underline{R}(a_1\hat{\underline{p}} - a_2\hat{\underline{b}}) - a_1\underline{u}(\underline{R})\hat{\underline{p}} \tag{8}$$

with $\quad \underline{u}(\underline{R})$ = plane vector of displacement,

so that (if the imaging parameters $\hat{\underline{p}}$, $\hat{\underline{b}}$ and $M(\underline{R})$ are known) $\underline{u}(\underline{r})$ can be obtained by approximation (i.e. multiple linear regression) from the $M(\underline{R})$ - values.
In most of the cases for $\hat{p} = \hat{b}$ in the undeformed state (load step k=0) the thinning-out coefficients a_i will be equal 1 .

2. IMAGE PROCESSING

The effective processing of the SEM-produced interference patterns presumes a computer aided organized handling in the steps

-digitization
-digital preprocessing
-discretization and
-evaluation .

The actuelly realized variant of MMT at The Institute of Mechanics is an off-line one using the image analysing system DENSITRON and a pretentious personal computer; an on-line variant which is just under work bases on the image processing system robotron A6472 .
In both cases the use of secondary interference patterns / 3 / produced by superposition of primary interference (fringe) patterns leads to alleviations in handling and interpretation:
If the patterns with

$$M_1^k(\underline{R}) = \underline{R}(\hat{\underline{p}}_1 - \hat{\underline{b}}_1) - \underline{u}^k(\underline{R})\hat{\underline{p}}_1$$

and

$$M_2^k(\underline{R}) = \underline{R}(\hat{\underline{p}}_2 - \hat{\underline{b}}_2) - \underline{u}^k(\underline{R})\hat{\underline{p}}_2$$

$(\underline{e}_{p1} - \underline{e}_{p2})$ at every load step k are combined the information about both components of displacement vector is given at the knot-points of a fringe mesh.

If the displacement of the mesh knots is measured for every load step $\underline{u}(\underline{R}^{k+1})$ can be determined as

$$u_i^{k+1}(\underline{R}^{k+1}) = u_i^k(R^k) + \left(M_i^o(\underline{r})\Big|_{\underline{r}=\underline{R}^{k+1}} - \dots \right.$$
$$\left. \dots M_i^o(\underline{r})\Big|_{\underline{r}=\underline{R}^k}\right) p_{i_k} \qquad (9)$$

so that the problem of fringe-order-calculation is objectless.

3. INTERPRETATION

The use of the measured displacements in the crack tip surroundings in fracture mechanical investigations is done in the step of interpretation and encloses the reconsideration of validity of special fracture concepts and the determination of critical values of characteristic quantity.

4. REFERENCES

/1/ Kühnert, R., Michel, B., Totzauer, W., Kämpfe, B., Hennicke, R.
WP G01 B/274 603/2

/2/ Morimoto, Y., Hayashi, T., Yamaguchi, N.
Strain Measurement by Scanning Moiré Method
Bulletin of JSME No. 233 27(1984)11, 2347-2352

/3/ Kühnert, R., Michel, B., Haberland, K.
WP G01 B/288 638/1

LINEAR FILTERING ON THE A6472 IMAGE PROCESSING SYSTEM BY USE OF RESIDUE ARITHMETICS

Luth, Wolfgang[1]; Gössel, Michael[1]

1. Introduction

The computation of weighted linear combinations of gray scale images or of different values in different pixels of a given gray scale image is crucial for such essential problems as linear finite response filtering, the resampling of images by bilinear interpolation, local moment computation, local orthogonal transformations, the accumulation of TV-frames and other.

In this paper we show how a weighted linear combination of gray scale images or of different values in different pixels of a given image can be computed fast and with high accuracy by use of the Robotron A6472 image processing system.

The computations are performed in residue arithmetics. The well known mixed radix convertion process is modified and adapted to the hardware possibilities of the Robotron A6472 image processing system.

2. The structure of the A6472 system

Since real parallel algorithms are always solutions of the problem how to compute on a concrete hardware with its advantages, restrictions and bottlenecks at first we will triefly describe the basic structure of the A6472 system. A detailed description is given in /1/.

The major components of the A6472 system are three independently programmable SIMD-pipelines or channels with four inputs and one output (s. Fig.1).

8 (or 4 in the case of the smaller configuration) image memories can be used as input and output memories of the channels. The capacity of an image memory is 512*512*8 bit. Instead an image memory a TV-camera can be directly connected to an input of a channel.

The memories are structured as a torus. Therefore a memory can be red with an arbitrary starting address (preset), but it is impossible to

---

1) Dipl.-Math. Wolfgang Luth, Dr. sc. Michael Gössel
Academy of Science of the GDR
Central Institute on Cybernetics and Information Processes
Kurstraße 33, Berlin 1086, GDR

read one memory with different arbitrary preset values simultaneous for different inputs. To weaken this restriction the so called "delay unit" can be used. It acts like preset values in X-direction such that some successive addresses in X-direction can be red and connected to different inputs of a channel. Into every data input and data output a free programmable look-up-table can be switched in to transform the coresponding input or output data stream by arbitrary unary functions.

A pass through the A6472 computer consists of the following steps:

a.) Programming of control data for the image memories, the arithmetical-logical units (ALU1,ALU2,ALU3), the input multiplexer and the normalization unit.

b.) Programming and activating of the look-up-tables EF1,EF2, EF3,EF4,AF.

c.) Sequential reading of data streams from the image memories, nonlinear transformations by the look-up-tables EF1,EF2,EF3,EF4, (8 to 10 bits), arithmetic-logical operations in the ALU1,ALU2, ALU3 (10 to 12 bits), diminishing of the word length in the normalization unit (12 to 10 bits), nonlinear transformation by the look-up-table AF (10 to 8 bits).

d.) Sequential writing of the output data into an activated image memory.

A complete pass through the A6472 system for a 512*512*8 bit image takes 80 ms.

3. Restricted word length and its implications

It is easy to see that a weighted linear combination of gray scale values can be computed by use of the look-up-tables and the addition operation of the corresponding ALU's.

Difficulties result from the restricted word length of 8 bits of the inputs and outputs of the channels.

a.) If the data and all results of multiplications and additions are restricted to a word length of 8 bits the accuracy of the results is not good enough.

b.) It is possible to use some image memories for the representation, but in this case we have to determine for every computation of a partial sum the corresponding carry bits and to add them in additional passes of the system.

It is a great advantage of the A6472 system that the multiplication of a data stream by a constant by use of a look-up-table and its addition to another weighted data stream can be computed in 80 ms for a 512*512 image. Therefore one should be careful not to waste passes for a simple carry addition if not nessesary.

c.) Since the A6472 computer consists of three independently programmable channels it is well adapted to residue arithmetics. The result can be computed in different rest classes without any carry and only once a conversation process to a positional number system is nessesary.

4. Basic notions of residue arithmetics

A residue number system is characterized by a set of L pairwise relatively prime integers $p_1, p_2, \ldots, p_L$.

For any integer z we denote $\langle z\rangle_{p_i}$ as the least positive residue with respect to p_i, i.e. $\langle z\rangle_{p_i}$ is the least positive integer q for which we have

$$z = m \cdot p_i + q, \qquad 0 \leq q < p_i$$

The interval (0,M) with $M = p_1 \cdot p_2 \cdot \ldots \cdot p_L - 1$ is called the legitimate range because for $z \in (0,M)$ the representation of z as

$$z \sim (\langle z\rangle_{p_1}, \ldots, \langle z\rangle_{p_L})$$

is unique.

The interval (0,M) can be divided into a positive and a negative region. The upper half of this interval represents the negative region, the lower half the positive one.

If only positive numbers are considered the whole interval can be used for the representation.

For z,z', $z \circ z' \in (0,M)$ we have

$$z \circ z' \quad (\langle\langle z\rangle_{p_1} \circ \langle z'\rangle_{p_1}\rangle_{p_1}, \ldots, \langle\langle z\rangle_{p_L} \circ \langle z'\rangle_{p_L}\rangle_{p_L})$$

where $\circ$ can be replaced by + (addition), - (subtraction) and * (multiplication).

I.e. the operations addition, subtraction and multiplication of integer numbers can be carried out componentwise without any dependencies between the different components.

After the computational process in a residue number system is finished there must be a residue-to-decimal conversation process.

a.) Chinese remainder theorem. z can be represented as

$$z = \sum_{i=1}^{L} a_i \langle z \rangle_{p_i} \mod p_1 \cdot p_2 \cdot \ldots \cdot p_L \,. \qquad (*)$$

with $a_i = p_1 \cdot p_2 \cdot \ldots \cdot p_{i-1} \cdot p_{i+1} \cdot \ldots \cdot p_L \cdot T_i$.

T_i is determined such that $\langle a_i \rangle_{p_i} = 1$.

The disadvantage of this method for the application om the A6472 system is the fact that the terms of the sum in (*) are relatively large and carry bits are to taken into account.

b.) Mixed radix conversation. z can be represented as

$$z = a_0 + p_1 \cdot a_1 + p_1 \cdot p_2 \cdot a_2 + \ldots + p_1 \cdot p_2 \cdot \ldots \cdot p_{L-1} \cdot a_{L-1} \qquad (**)$$

with $p_1 < p_2 < p_3 \ldots < p_L$, $\quad a_0 < p_1$, $a_1 < p_2$, ..., $a_{L-1} < p_L$.

The coefficients $a_0, a_1, a_2, \ldots$ can be recursively determined by

$$a_0 = \langle z \rangle_{p_1}$$

$$a_1 = \langle (\langle z \rangle_{p_2} - a_0) \cdot q \rangle_{p_2} \,, \qquad \text{with } \langle q \cdot p_1 \rangle_{p_2} = 1 \,.$$

$\vdots$

A direct application of the mixed radix conversation process has the disadvantage that the least significant 8 bits and the most significant 8 bits of the result in positional representation are not separated. Therefore a modified mixed radix conversation process has to be used.

5. Weighted linear combinations with a resulting 16-bit word length

Let be

$$S = \sum_{i=1}^{N} a_i \, x(i)$$

where x(i), $0 \leq x(i) \leq 255$ is a nonnegative gray scale value.
If we have for all i $a_i \geq 0$, then we suppose

$$\sum_{i=1}^{N} a_i \leq 255.$$

This supposition quarantees us that we have $0 \leq S \leq 65025$.

In the general case, if the coefficients a_i could be positive or negative we suppose

$$\sum_{a_i>0} a_i \leq 127 \qquad \text{and} \qquad \sum_{a_i<0} -a_i \leq 127$$

and we have $-32385 \leq S \leq 32385$.

The computation of the weighted sum on the A6472 computer will be simultaneously carried out within two pipeline channels modulo 255 and modulo 256 respectively.

The programmable look-up-table EF1 of the first and the second channel is loaded with

$$f(x) = \langle a \cdot x \rangle_{255} \quad \text{or} \quad f(x) = \langle a \cdot x \rangle_{256} \qquad \text{for } 0 \leq x \leq 255$$

respectively.

The new incoming data and the previously accumulated sum are added. The programmable look-up-table AF of the first and the second channel is loaded with

$$g(x) = \langle x \rangle_{255} \quad \text{or} \quad g(x) = \langle x \rangle_{256} \qquad \text{for } 0 \leq x \leq 1023$$

The conversation process to a positional number system relies on the representation

$$S = \langle S \rangle_{256} + 256 \cdot A_1$$

$$\text{with } A_1 = \langle \langle S \rangle_{255} - \langle S \rangle_{256} \rangle_{255}.$$

The eight least significant bits of S are $\langle S \rangle_{256}$ and the eight most significant bits of S are A_1.

If all a_i are nonnegative the final result is computed. If the coefficients a_i are positive and negative we have to determine the sign of the result yet.

Then we have

$$S_V = \begin{cases} S & \text{for } 0 \leq S \leq 32639 \\ S - 255 \cdot 256 & \text{for } 32640 \leq S \leq 65279 \end{cases}$$

and since $|S_V| \leq 32385$

$$S_V = \begin{cases} S & \text{for } 0 \leq S \leq 32385 \\ S - 255 \cdot 256 & \text{for } 32895 \leq S \leq 65279. \end{cases}$$

and also

$$S_V = \begin{cases} S & \text{for } A_1 \leq 126 \\ S - 255 \cdot 256 & \text{for } A_1 \geq 128. \end{cases}$$

For a representation in the 16 bit two's complement we get finally

$$S_{v2} = \begin{cases} S & \text{for } A_1 \leq 126 \\ S+256 = 256(A_1+1)+\langle S\rangle_{256} & \text{for } A_1 \geq 128 \end{cases}$$

The conversion process can be realised in both cases in one pass of the system.

6. Examples and computing time

In one channel of the A6472 computer up to 3 terms of the sum can be added to the till now accumulated partial sum. The computation modulo 255 and modulo 256 can be carried out simultaneously in two channels of the system. But the possibilities of such parallel work depend on the number of available image memories.
Table 1 shows the number of passes in dependence on the number N of terms of the sum and the number of available image memories for the realization of an FIR-filter. The results are compared with the direct method, where carry bits are to taken into account. The advantage of the residue method is obvious.

number of image mem.	direct method	residue arithmetics
4	---	2N+1
5	2·N	N+1
6	$2\frac{N+1}{2}+1$	$\frac{N+1}{2}+2$
7	$2\frac{N+1}{2}+1$	$\frac{N+2}{3}+2$
8	$\frac{N+1}{2}+1$	$\frac{N+2}{3}+2$

Table 1: Time complexity of a FIR-filter on the A6472 system

Table 2 shows the computation time for the following four examples:

Example 1

Realization of a FIR-filter with arbitrary coefficients and a filter mask (N=49 points) of arbitrary form

Example 2

Computation of the arithmetic mean of the gray scale values in a quadratic window of size 15*15 (N=225 points)

Example 3

Translation of a gray scale function by a vector r=(,), r 1. Bilinear interpolation is used to resample the function (N=4)

Example 4

Accumulation of TV-frames (N=20). The processing speed is restricted by the frequency TV-frames can be red into the system (f=12,5Hz)

Example	N	M=4	M=7
1	49	8	1.52
2	225	5	1.52
3	4	0.24	0.24
4	20	1.68	1.68

Table 2 Computation time (sec.)

Literature

/1/ Kempe V. et al.: Interaktives Bildverarbeitungssystem robotron A6472; "Neue Technik im Büro", 26(1982)5 131-137

/2/ Szabo N. S.; Tanaka R. I.: Residue Arithmetics and its Application to Computer Technology; New York McGraw Hill, 1967

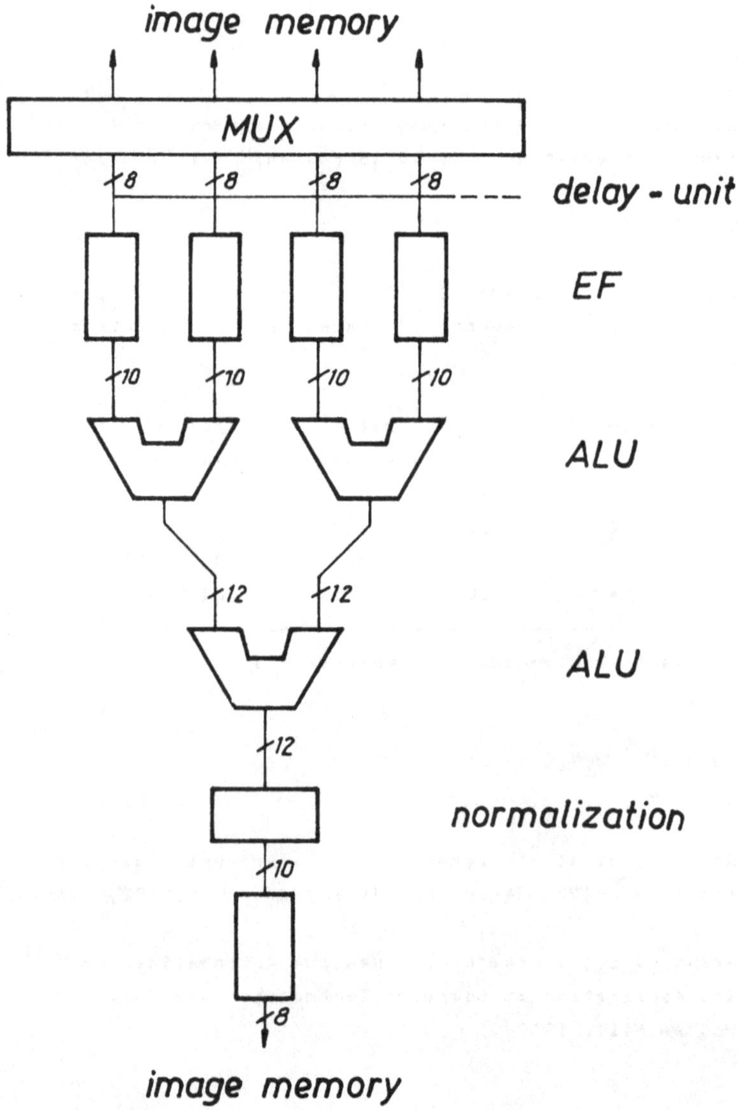

Fig. 1: Scheme of one of the channels of the A6472 system

PROGRAM PACKAGE FOR PROCESSING LINE FLOW LIKE IMAGES BY BVS A6472 USING ADAPTIVE AND ANISOTROPIC METHODS

Unger, S., H.Winter, Chr.Feist, U.Marzok, L.Zedler [1])

Abstract

Line flow images may be defined images that local line direction and line spacing vary smoothly, while the critical points of line flow being sparsely distributed over the image. Examples are optical interference patterns, iso-density curves on radiographs etc.
By our approach is intended to detect line segments, that follow the local line flow direction, with considerable robustness against medium sized disturbations (of random nature as also irrelevant deterministic substructures (holes,patches)), shading, and noise introduced by the images physical transmission processes.
The main tools to reach that goal by low processing afford comprise:
- preprocessing by logical operators of several sizes and a rule for combining the resulting images - intended to reduce the shading and noise disturbances,
- detection of the local line flow direction to mark critical points of the line flow, and of empty regions to exclude them from the anisotropic processing steps, resp. all further processing at all.
- use of the detected line flow direction image to control the succeeding line extraction steps (anisotropic processing)- intended to suppress the medium sized disturbances.
- Sceleton restoration , using else direction information to find the right point pairs to be connected (to close line breaks), and to eliminate apparent arcs, loops etc.
- Contour processing alternatively to thinning and sceleton processing for feature extraction

1. Introduction

The main steps of the approach are as follows:
- pre-processing for line- and line direction detection, based on criteria for the combination of images filtered by Pseudo-Laplacian operators [1,2] of sizes from a fixed, prespecified ordered set. A thresholding parameter, dependent on prior estimated noise amplitudes, serves to reduce noisy degradations. By the type of logical filtering used, an efficient shading correction is warranted. The method works quite well in the presence of local varying line spacing and thickness parameters.
- after pre-processing, the line direction image is detected by fitting the level crossings of intensities by 3 by 3 masks, followed by smoothing. The resulting image would be used to control the succeeding line detection step [3]. By that method the lines are detected with considerable high robustness against structural disturbations (holes in lines), and spatial correlated disturbations (dark/bright patches in regions of high/low intensity), except the loci of critical points (comp.4.)
- use of local line directions to control several repairing (restoring) steps on sceletons.

The requirements of a short main run time constrained the choice of image preprocessing methods to be appropriate for implementation and fast execution by the available computing hardware (5.)

The line detection method implemented comprises:
- an intensity transforming step to achieve a position-invariant scale

[1]) Akademie der Wissenschaften der DDR, Zentralinstitut f. Kybernetik und Informationsprozesse, Kurstr.33, 1086 Berlin

of the typical local line intensity variations (2.), - followed by the detection step of local line flow direction (3.), which controles the operations of
- anisotropic smoothing (5.).
Binarization then can be done by an universally definable threshold. The succeeding anisotropic restoration and thinning steps represent slightly modifications of convenient procedures to avoid accidental bridging of lines and shortening of line endings.

2. Line-size-adaptive, isotropic image enhancing

As mentioned above, by the filtering method used, operators of various sizes are combined by a rule. Similar procedures are known from edge detection [4], [5].
From reasons of simplicity, every contributing fixed-sized operator had been based on Pseudo-Laplaceian filtering, by which only the signs of intensity differences between the central pixel P, to be updated, and those of a surrounding sampling subset $\Omega(P)$ of image pixels are processed. The set $\Omega(P)$ should meet the following conditions:

1) near rotational invariance

2) the elements of the set $\{\Omega(P,1),\ldots,\Omega(P,k)\}$ of sampling sets, generated by a series of k values of operator sizes should be pairwise disjunct - to achieve maximum sensitivity on the operators sizes.

Evidently, the conditions 1) and 2) would be fulfilled by ring-shaped sets $\Omega(P,r)$, with r being the radius of the respective concentric circles, together with a set of values r_1 to r_k, selected by prior knowledge about certain properties of a class of images (see below).

The fixed-sized operator is closely related to the empirical distribution function $F_{emp}(v,\mathfrak{F}(P,r))$ - that is the frequency of intensity values on $\Omega(P,r)$ being less than v, $v \in R^1$. By $\mathfrak{F}(P,r)$ is denoted the corresponding ordered set of sampled intensities from $\Omega(P,r)$. The operator then can be defined by the parameters

- r (filter size)
- s:== threshold for intensity differences to be valued significant relative to a known noise amplitude of the physical image's sampling and transduction process, and its updating function

$$g(P,r) = Q*(F_{emp}(u-s,\mathfrak{F}(P,r)) - (1-F_{emp}(u+s,\mathfrak{F}(P,r)))), \quad (1)$$

where $F_{emp}(-\infty)=0$, $F_{emp}(+\infty)=1$,
$g(P,r) \in [-Q,Q]$,
$u=f(P)$:== intensity value in P,

Q:== amplitude of the filtered image.

An output of the fixed-sized filter equal to zero - being the center of the filters scale of transformed intensities - marks the threshold in classifying a region as a member of classes 'ridge' or 'valley', the reject region.

Some properties of fixed-sized operators, which had been observed in simulation, are shown in fig.1 and fig.2. They contain one-dimensional intensity profiles of four continuous input images and the profiles of the results for three sizes of ring-shaped filters (nearly homogenously dense sampled on $\Omega(P,r)$ and two cases of threshold s. The input image in fig.1a) is an ideal periodical two-dimensional sinusoidal intensity image having unit wave-lenght d. The intensity does not depend on the second dimension. Fig.1d) is a shaded version of a). In fig. 2a) and d) a single circular local depression of ridge intensity is added. (The shown profiles go through the center of depression.)

The optimum radius r_o of Ω for sampling intensities inside a 'local region' of the image may be defined - partly intuitively - as follows:

- Let $h(P,r) := \sup\tilde{g}(P,r) - \inf\tilde{g}(P,r)$,
  the range of sampled intensities on $\Omega(P,r)$, and
- Γ a fixed 'local region of diameter d', containing at least points P_1 and P_2
  with P_1 on 'ridge' and P_2 on 'valley', and let
  $w = f(P_1)-f(P_2)$ be its corresponding intensity differences

then $\quad r_0 = \inf \{ r : \forall P \in \Gamma \; (h(P,r)=w) \}$,

that means, that $\Omega(r_0)$ is the smallest-sized set sampling the local maximum and minimum intensities for all image points from the local region itself - in the case of a sinusodial unshaded intensity function $f(x,y)$ the optimum value r_0 becomes equal to d/2 (d being the wave length of f in this case).
The rather important feature of the fixed-sized operator is the dependence of its output characteristics on the relative filter size r/d in the case $r/d \gg 1$:
Because of the relative large diameter of the set Ω the range of sampled intensities on Ω is sensitive to shading, that is, will be streched by linearly varying contrast. By (1) this leads to a corresponding decrease of the absolute values of $g(P,r)$, resp. shrinking of the ridge/valley discriminating intensity differences, after having applied the operator, compared to filtering by an operator size equal to r_0. Consequently, the absolute value of the output of a too large sized operator is depressed.

The whole procedure only works well, if the lines can be detected from 'local' intensity shapes. Likewise is a decision on the significance of any line-like intensity profiles inside ridges or valles based on their respective prominance relative to the intensity profiles in the immediate neigbhorhood. Naturally, never may be expected, that any intensity distribution, which visually may be line-like, would be correctly detected by such a procedure.
In the absence of shading and parallel, equally shaped and d-spaced straight lines, the dependence of $g(P,r)$ on the size r for any fixed position P becomes small in the limit of $r/d \gg 1$, such that the dependence of $g(P,r)$ across the lines in any direction else becomes nearly independent on filter size above r/d=4, say.
On the contrary, the shapes of $g(P,r)$ in the case of small relative filter sizes are different and dependent on r, comp. the curves in fig.1 with r/d=0.5 and r/d=0.125.

A good value for threshold s has some agreeable effects:
- The treshold s, if adjusted slightly above the amplitude of exepted noise or other disturbances, serves to map homogenous intensity regions to values equal to (or near by) zero and supresses the detection of apparent structures (comp. fig.2f); the local depression of ridge does not influence the output of filter with r/d=0.125).
- The output of too large and too small filters is much more depressed than the output of the optimal filter (comp. fig.1c) and f)).
- Shading-effects can be reduced (comp. fig.1f)).

This leads to the simplest combining rule

$$g(P) = g(P,r_1), \text{ if } |g(P,r_1)| \geq |g(P,r_2)|$$
$$g(P) = g(p,r_2), \text{ otherwise} \qquad (2)$$
$g(P)$:== the resulting output of combined operators of sizes r_1 and r_2

for the attenuation of the filter size to the local optimum size $r_0(P)$ from above. To achieve that attenuation, the series of filter sizes r_1 to r_k has to be chosen suitably from knowledge about the class of images to be processed - as mentioned above:

r_1 resp. r_2 :== the maximum resp. minimum line spacing parameters d to be expected in the class of images resp. in the same image side by side.

A more complicated combining rule to reduce disturbances like in fig.2 uses a threshold t to favour the bigger filter size, if different signs occure:

$$g(P) = g(P,r_1), \text{ if } (\mathrm{sign}(g(P,r_1)) = \mathrm{sign}(g(P,r_2)) \text{ and } |g(P,r_1)| \geq |g(P,r_2)|) \text{ or } (\mathrm{sign}(g(P,r_1)) \neq \mathrm{sign}(g(P,r_2)) \text{ and } |g(P,r_1)| \geq t \text{ and } r_1 \geq r_2)$$

$$g(P) = g(p,r_2), \text{ otherwise.}$$

3. Detecting the local direction of unoriented lines

The angle α of unoriented directions is defined $\in [0,\pi)$ and can be calculated from corresponding oriented directions by the $\mathrm{mod} \pm n\pi$ operation, n=1,2,..

Detection of the α image has been based on first derivatives of the two-dimensional intensity image $f(x,y)$:

Let $q(x,y)=|q|(\cos\beta,\sin\beta)$ an estimate of the local intensity gradient, then

$$\tilde{\alpha} = \beta + \pi/2 \ \mathrm{mod} \pm n\pi, \quad n=1,2,..,$$

is an estimate of the local line direction. The derivatives generally are ill defined, therefore the $\tilde{\alpha}$-image has to be extensively smoothed, which is permitted from its intrinsic large scale property as mentioned above. For that, linear filtering had been applied to the image

$$e(x,y)= h\cdot(\cos 2\tilde{\alpha}(x,y), \sin 2\tilde{\alpha}(x,y)),$$

where $h=1$, if $|q|>0$, $h=0$ otherwise,

resulting in a smoothed image

$$E(x,y)=|E(x,y)|\cdot(\cos 2\alpha,\sin 2\alpha).$$

$|q|$ may become zero, if

- the position lies inside an indicator set of a constant intensity level, e.g. a value from {0,1,2,..,15}
- the estimator itself being undefined (e.g. no mask of a set of 3 by 3 masks does match)

By sequentially applying a sufficient number of smoothing steps, all undefined points become defined ones.

4. Detecting Critical Points

Critical points of line flow can be detected by turning counter-clockwise around the points of the line direction image along circles. Therefrom that points can be classified according to the sign (s) of angle changes and the number q of passed $[0,\pi)$-intervals during run through the full $[0,2\pi)$-turn (see fig.3):

line flow characteristic	s	n
circular	+	2
parabolic	+	1
hyperbolic	−	>1

All other points are classified regular image points. A zone around every critical point can be withdrawn from anisotropic line processing steps (3.2).

5. Line detection

After having executed the image enhancing step (2.), the transformed image intensities on ridge and valley have near constant positive/negative values throughout the image such, that shading had been removed. Binarization then can be done by a single threshold.

Alternatively, the transformed image may be first improved by anisotropic smoothing, that is, by local application of directional smoothing filters for different filter-angles, dependent on the prior detected local line direction α - and then can be binarized using a single threshold, which can be easy calculated from the filters parameters.

6. Feature extraction (for line restoration)

It may be performed by sceleton processing or, alternatively, by processing the line contours - avoiding thinning.

Useful features for restoring lines from sceletons are:

First order attributes:
- type of points: 'endings'(1), 'interior line points'(2), 'bifurcations'(3), 'crossings'(4)
- coordinates (x,y) of the points of special type

Set labels:
- label K of the distinct one-element-sets (e.g. points of type 1) to 4) with respect to their video scanning order)
- labels L of the distinct sets of connected points of type 2

Second order attributes:
- an ordered tripel of numbers L, each being incident on a point of number K (comp. fig.4 for type 1 points)
- incidence relations between all labelled sets
- oriented direction of L-labelled lines, being incident on type 1, resp. type 2 to type 4 points

Preprocessing for calculating features from contours:
The processing of contours - to avoid thinning - is based on a parallel (position-invariant and non-recursive) computing of counter-clock and with-clock oriented secants between each point of a contour and its n-th predecessor/ successor along the contour (comp. fig.5). Both secants images may be stored in the same image memory (for saving of storing space) dependent on the wanted range of numbers n, but must not. The main steps comprise:
1) checking the contours, if they being free from double points. If the check fails, the respective loci will be broadened by extra pixels for the storage of secant information.
2) signing of the contours pixels (which form a closed linear chain, of positions 1,2,..,l,..)
3) generating images A1(l) resp. A2(l), which contain the relative position codes of the first predeceesor (l-1) resp. the first successor (l+1) contour point relative to the l-th point. A1 and A2 are defined for every point (l) on the contours, otherwise undefined. A1 and A2 would be generated simply from look up tables on the respective 8-bit-coded 3 by 3 windows.
4) generating the four images E1X(l,1),E1Y(l,1),E2X(l,1),E2Y(l,1),which contain the differences of the x- resp. y-coordinates between the points l and l-1 resp. l and l+1. EjX, EjY can be defined by look up tables from Aj.
5) calculating the respective images Ejz(l,n), j=1,2, z=x,y. These contain the coordinate differences between the l-th and l+n-th, resp. l-th and l-n-th points along the contour. The calculation may be done by the following recursions ('rotations of the contour')

$$Ejz(l,n+1)=Ejz(l\pm1,n)+Ejz(l,1),\ n=2,3,..,\ \text{'+' for } j=2,\ \text{'-' for } j=1. \quad (5)$$

That calculation is done by shifting the image 8 times, and executing (5) if the pixel l+1 resp. l-1 coincides with the pixel l, what can be checked from coincidences of countour markings generated by step 2).

From the secants the curvature and its sign (concave/convex) and therefrom other features (line endings bifurcations, if any) may be recognized (comp. also [6]).

7. Implementation

The image processing system used is a robotron A6472. The video processins device of the system has the following main characteristics:

- it is suited for the implementation of algorithms of non-recursive position-invariant type
- the intensities of both, input and output images, are values on an 8 bit scale
- three 12-bit binary addition/subtraction pipelined arithmetic devices work in parallel, their operating modes being adjustable by suitably programming the controller of the video processor, including logical functions too
- eight 8 bit image intensity memories, each of 512 by 512 pixels, are available by programming
- there are means to transform the input intensities by looking up tables and to spatially shift the whole image an arbitrary number of rows and columns, where the resulting positions become defined mod 512. Every such transform can be selected by programming the controller too.

There is an interface for the exchange of data between the image memories and the central memory af a general purpose host computer. It is mainly destined for gathering image features from small subsets of significant sceleton/edge points which - after preprocessing - contain the relevant information of the image.

All computationally extensive preprocessing steps had been implemented on the video processing device. Only sceleton restoration is done by the host computer but uses the video processing device for some provisional results too.

Several versions of ring-shaped Pseudo-Laplace-operators are implemented. For the reason of further reducing the influence of noise and disturbances, not only intensity differences between the single central pixel P and those of surrounding sampling set can be used but also differences between a central sampling set and a surrounding one. Run time depends on the number of those differences. It can be reduced by using only a subset of the digital circle (which becomes important for big values of r) and by using pixel pairs for intensity differences so that the number of vectors becomes small. Fig. 6 shows an example where all 32 pixel pairs are defined by two parameters A and B. The run time for this operator is 3.01 sec. In general (2+3n/2)*0.08 sec is needed (n is the number of intensity differences).

The run times for other steps are (dependent on parameters):
- detecting the local direction (incl. smoothing) 3-6 sec
- anisotropic smoothing 5-8 sec
- thinning 30-60 sec (dependent on the image)
- sceleton restoration 30-60 sec (dependent on the image).

The experimental results for optical interference patterns proved the efficiency of our method.

References

[1] M.Nadler: Effective and cost-effective real time picture operators for medical imagery. In: F.T.Dombal, F.Gremy: Decision making and medical care. North-Holland, Amsterdam 1976

[2] Letellier, Nadler, and Abramatic: The Telesign Project. Proc. of the IEEE, 73(1985) 4, p.813

[3] H.Knutson, R.Wilson, P.H.Granlund: Content dependent Filtering of Images. Proc. of the Int. Conf. on Digital Signal Processing, Florence, Italy, Sept. 2-5, 1981

[4] A.Korn, C.Erdtel: Kombination verschiedener Filterkanaele zur Optimierung einer Merkmalsrepraesentation im Bildbereich. 7.DAGM-Symposium Erlangen, September 1985

[5] V.Torre and Tomaso Poggio: On edge detection. PAMI-8, No.2, p.147, March 1986

[6] J.Saedler, W.Wilhelmi, V.Kovalewski, N.Eichhorn: Implementierte Algorithmen auf dem Displayprozessor K 2027 (II). Bild und Ton 39 (1986) 6, S. 165.

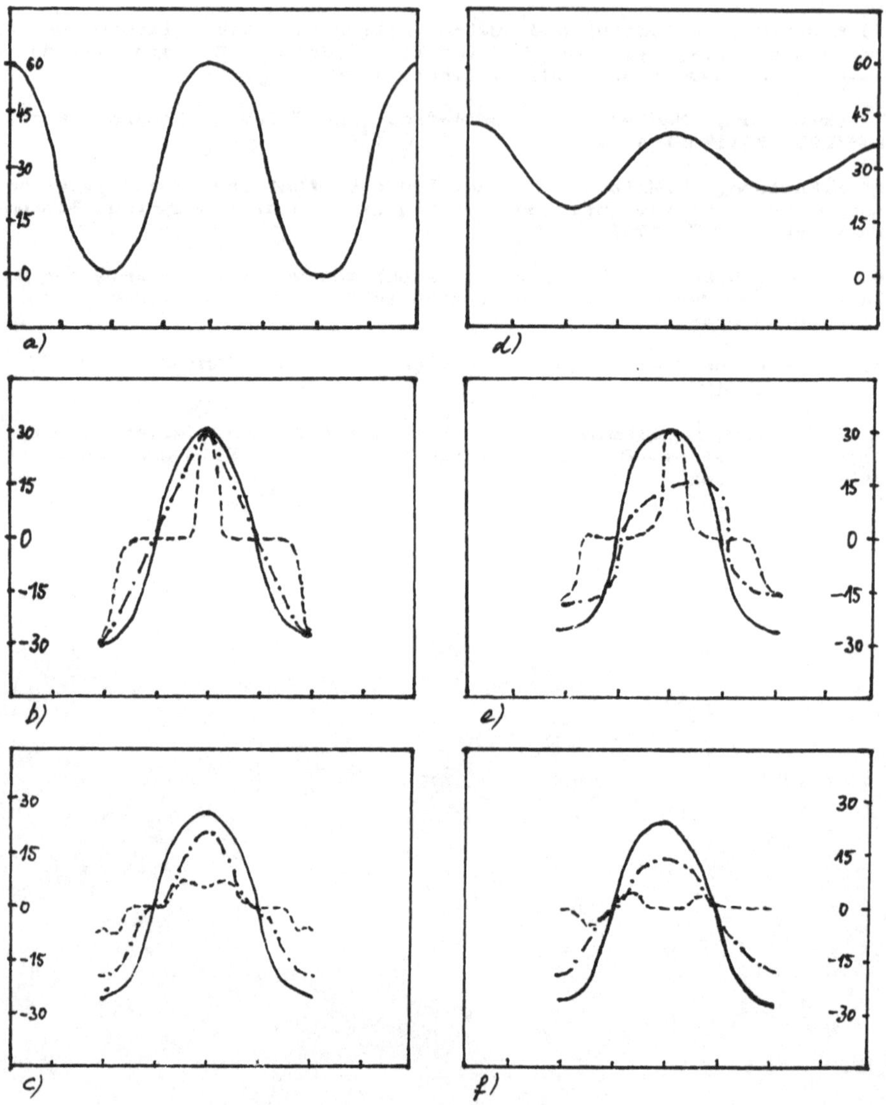

a), d) intensity profile of input images

b), c) operator output for s = 0
c), f) operator output for s = 8

relative filter size:
–·–·– r/d = 4
——— r/d = 0.5
- - - - r/d = 0.125

fig. 1   Properties of fixed-sized operators

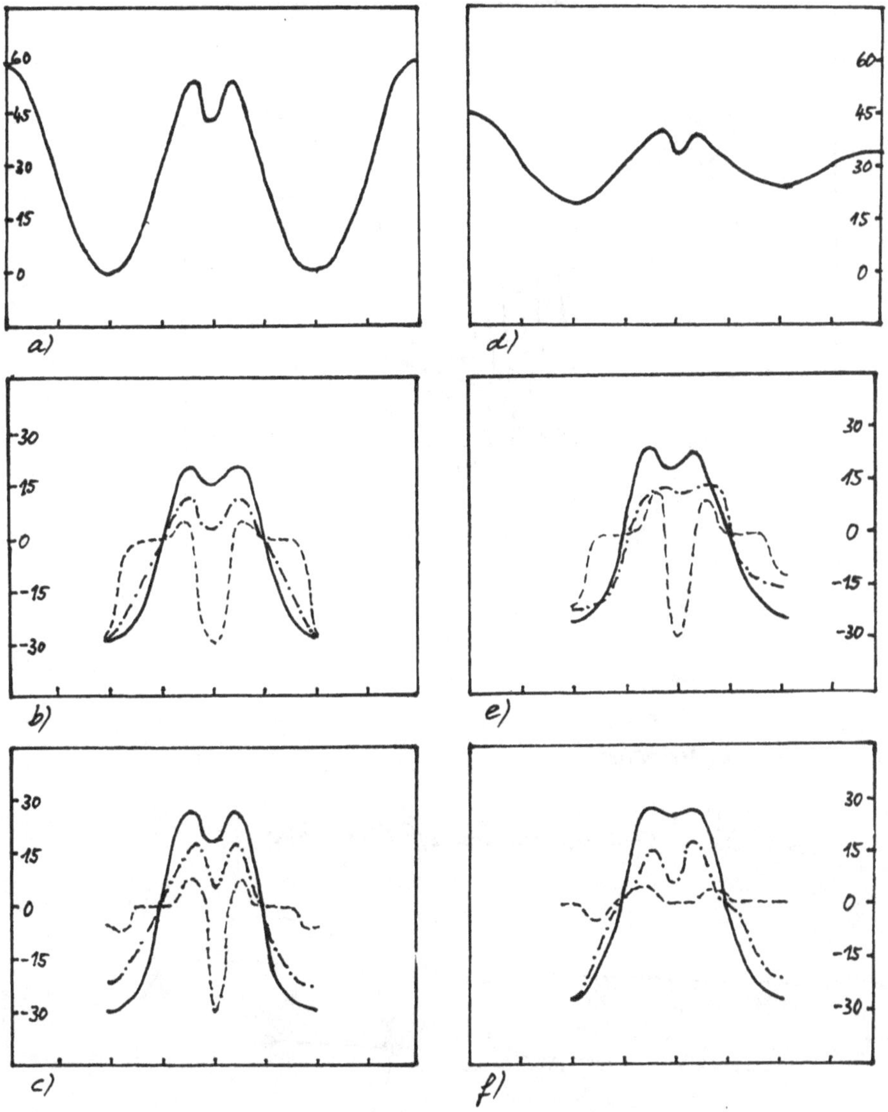

a), d) intensity profile of input images

b), e) operator output for $s = 0$

c), f) operator output for $s = 8$

relative filter size:

— · — $r/d = 4$

——— $r/d = 0.5$

- - - $r/d = 0.125$

fig. 2 Properties of fixed-sized operators

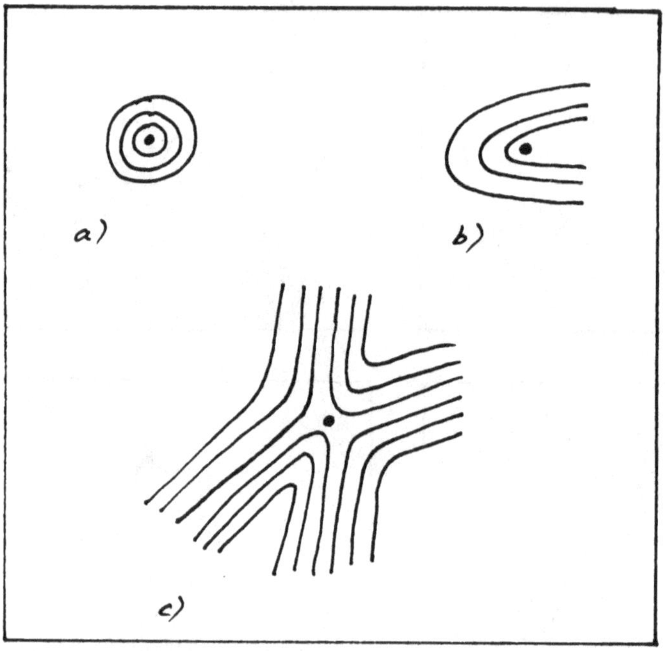

a) circular
b) parabolic
c) hyperbolic

fig. 3 Critical points of line flow

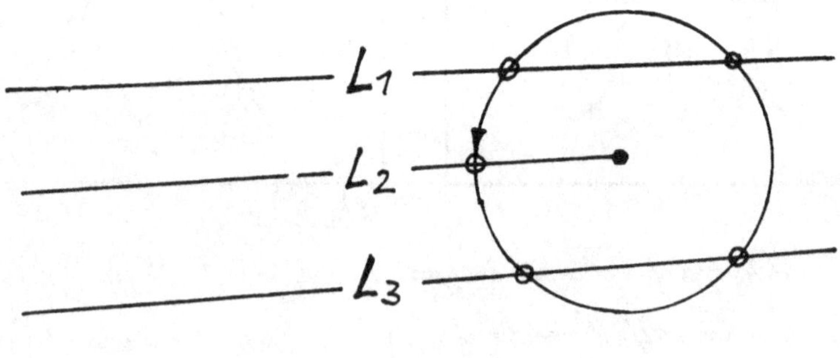

fig. 4

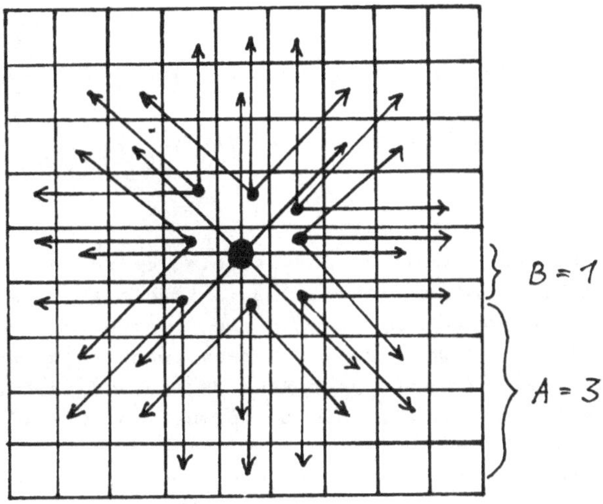

fig. 6 A Pseudo-Laplace-operator

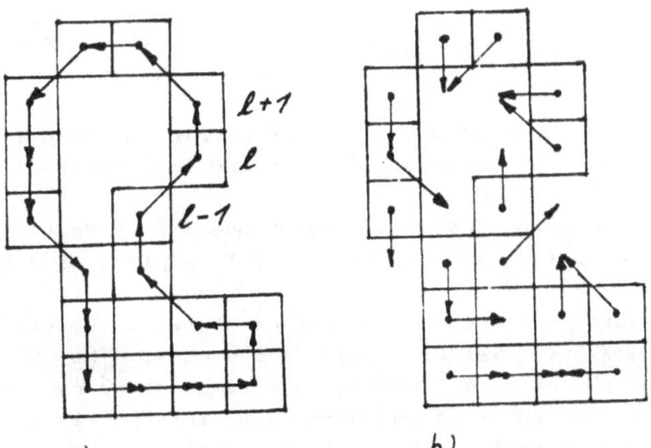

a) image of relative position-vectors to successors, $n=1$

b) the same image rotated clock-wise, $n=1$

fig. 5

THE PROGRAMMABLE HARDWARE-STRUCTURE GIPP AND ITS APPLICATION

Kutschke, G. and G. Schwarze [1)]

1. Introduction

Considerable progress has been achieved in designing efficient processor architectures for detecting edges or contours in images and identifying them with real ones in the scene mapped onto the image plane. The parallelism inherent to most of image processing steps leads in a straightforward manner to parallelism in the processor design and in this line to a high repetition rate in a single chip or to multi-usable IC's for a single processor.

In the case of using local operations in image processing it seems to be appropriate to distinguish between parallelism at the entire image level and at the processing window level. For the latter case we will give an example for an efficient high performance structure. The base for this structure is the microprogrammable special purpose processor GIPP (GRAYSCALE IMAGE PREPROCESSOR), developed at ZKI.

2. The processing algorithm of the GIPP and the resulting hardware-structure

The objects in a picture are often distinguished by regions of different graylevels. Different image processing functions are components of recognition procedures for such objects.

The GIPP realizes these image processing functions in a raster-scan approach with a characteristic movement of the central pixel of a window (Fig. 1).

In general the computation of the new central pixel value consists in selecting one of the graylevel values inside the window. The selecting rule is determined by the given processing function. Masks can be set. If the mask is true, the computation and setting of the central pixel is carried out, if not, the central pixel remaines unchanged.

Generally, the GIPP works in a bit-plane manner. One iteration step is needed for every bit plane. In every iteration step the available graylevel range G is divided by two into a lower greylevel range G_e and a upper graylevel range G_u .

By using the select-rule it is tested, which graylevel range contains the exspected value. The graylevel range computed in the actual step is the available graylevel range G in the next iteration step.

1) Central Institute of Cybernetics and Information Processing of the Academy of Sciences of the GDR (ZKI) Kurstrasse 33, Berlin, 1086 GDR

During every iteration step the graylevels of all pixels of the window are compared with the actual middle graylevel g_m . Every comparator generates a variable w_i . The nine results w_i of comparison in the window are variables of a function F which realizes the image processing operator.

The value of the function F in every iteration step determines the computation in the lower or upper graylevel range for the following iteration step. After m=log NG iteration steps the new graylevel is computed (NG=number of graylevels). /1/

The realization of the described algorithm by means of a hardware structure is shown in fig. 2. The schematic circuit contains the following main parts:

- 3x3 register-block

  In the 3x3 register array the GIPP records the graylevels of 3 consecutive pixels in corresponding positions of 3 consecutive lines G(0)...G(8). Every input action moves the window one step along the lines of the picture, line by line.
- 3x3 comparator-block

  The nine graylevels of the window are compared in a 3x3 comparator array with the contents of the register E g_m in every iteration step. The outputs of the comparators are identical with the nine variables of the function F.
- table memory

  The function F which decides the implemented picture processing function is typically generated by reading a 512x1 bit RAM.
- The controller has two tasks:

  1. At the beginning of every iteration step the controller sets the register E on the half value of the available graylevel range g_m.

  2. At the end of every iteration step the actual bit plane of register E is set or reset in dependence of the value of F.
- register E (result register)

  The register E is a temporary memory for the comparation-value g_m. At the end of computation the register E contains the result.

A simple extension of the GIPP-hardware-structur makes it possible to carry out table-lookup operations too. By this way Freeman chain coding combined with node coding can be accelerated.

The described hardware-structure GIPP was realized by standard-IC's on a printed card and is operational. The realized GIPP-boards are an integral part of the picture processing system BAS4000. The application of such system is described elsewhere. /2/

3. Survey about the implementable image processing functions

The basic idea of the processor GIPP is the effective hardware-implementation of a specialized processing algorithm described in the foregoing chapter by optimization of a parallel working and programmable hardware-architecture. Therefore it is possible to implement all image

processing functions in the processor GIPP which can be mapped directly or in a decompositional form onto this processing algorithm.
These image processing functions controlled by microprogramms are for instance:

- mask operations as thinning, thickening, restoration, directional median
- sorting operation as rank-filter, including median, maximum- and minimum operations
- table-lookup-operations as data compression by Freeman-chain coding, combined with node coding
- generation of all structuring elements suited for image algebra operations (in the sense of Matheron, Serra and Sternberg) as dilatation, erosion, opening, closing et.al.

4. The schematic circuit of a GIPP, aspects of realization as semicustom chip

The GIPP-algorithm may be characterized by four essential properties:

- Pictures are processed by moving a 3x3 window in raster-scan manner.
- Within every bit plane the GIPP works in parallel.
- The GIPP computes the several bits of the result in a serial manner (bit plane processing).
- The number of needed computation steps is constant and depends only from the given number of graylevels NG.

Because of this properties the GIPP has a high repitition rate of internal substructures and is characterized by a synchronous processing modus which realizes a constant processing time for every pixel, independent of the pixel value.
Fig. 3 shows the schematic circuit of a GIPP for realizing as semicustom chip. Nearly all depicted function blocks are characterized by a high repitition rate of substructures. For example the identical function block RCU is contained nine times in the shown circuit. The types, the number and the arrangement of its identical internal substructures verify that the GIPP-structure is very well suited for realization as semicustom chip in the shown manner.

Short description of the schematic circuit (Fig. 3):

The main data lines, shown in fig.2., are accentuated. The 3x3 register-comparator-block is realized by the nine register/comparator units RCU0...RCU8. The input lines DI10...DI37 deliver the values of the corresponding pixels of three consecutive lines to the registers. The nine graylevel values of the window, stored in the registers, are compared with the contents of the result register RR E0...7, transmitted by the comparator controller CC.
The nine outputs w of the comparators are suited with the address pins

ATØ...AT8 for the external table memory. The selected memory contents is transferred to the result-register-controller RRC by the lines TSV and TM. RRC sets or resets the several bits of the result register RR in the described manner. The contents of RR EØ...7 is transferred to CC to generate the new value of comparison for the next step. At the end of computation RR contains the result which is transferred to the output lines DOØ...7.

Besides this main function the schematic circuit includes a set of programmable processing functions and parameters which allows a high adaptability to given applications. The most important programmable attributes are:

- adaptable number of graylevels by a programmable length of the clock shift register CSR
- different function modes, for instance programming mode for the external table memory,mask and non mask operation modes, border line detection mode, Freeman-code-mode
- different modes for GIPP-coupling: single GIPP- and pipeline mode, high speed-coupling
- programmable threshold for Freeman-code generation
- table memory selection for connection of longer table memories which contain a set of tables
- counting of the actual line position and signalisation of the programmable line end for Freeman-code generation.

The position- and address-counter PAC generates the line position ZPØ...11 for Freeman-code generation and the addresses AZØ...11 for the connection of a line memory shown in fig.4.

In a special mode, programmable via the line BA, it is possible to use the complex "line end controller ZEC and PAC" as a separate unit. A useful application of this possibility results from the Freeman-code generation in the mode "high speed coupling".

The high adaptibility is a decisive property of a integrated picture processing circuit for general applications.

Fig. 4 shows the principle connections of the line memory, of the table memory and of the clock generator with the GIPP. All needed address select and control signals are generated by the GIPP. The expense of external elements is minimized. The computation speed in this single GIPP application is decided by the computation time for every bit plane, that means by the technological based delay times for the GIPP and the access time of the table memory. For the proposed GIPP realization as semicustom chip the typical computation time per bit/plane amounts to 200ns. To compute 8 bit pixels about 9x200ns=1.8µs are needed.

5. GIPP-networks for a new class of power properties and image processing functions

5.1 Pipelining of GIPPs

Fig.5 shows the pipelining of n GIPPs as a MISD structure for simultanous performing n different image processing functions. Each GIPP is programmed by loading a microprogramm for one of these n functions. The synchronous processing mode of the GIPP allows this architecture without severe control problems. The overall processing time is approximately equal to the processing time in the case of one GIPP and one processing function.

5.2 GIPP arrangement for advanced processing speed

For high speed computation a cyclic arrangement of GIPPs is developed (Fig.6). The consecutive GIPPs compute consecutive pixels of the result picture. The cyclic arrangement must include such a number of GIPPs, that the activation of a GIPP for a new computation may occur only after closing the computation of the preceding pixel by this GIPP. On the strength of this principle the maximum computation speed is decided only by the possible input and output speed of picture data, that is by the level of the technology. On the strength of actual technology the highest possible computation speed is limited to 10 Mc/sec.

5.3 Parallel-pipeline architectures of GIPPs

A practicable solution with a wide application field consists of a structure with m≥2 parallel pipelines and n GIPPs in each pipeline. The m pipelines are connected after each stage of the n GIPPs with an interconnection network. In this network simple arithmetic, logical and shift operations may be implemented . The main goal of this architecture is the implementation of all structuring elements of the image algebra.

References:

/1/ Roesler, U., G. Schwarze, T.L. Chung: Lokale Bildoperatoren zur Manipulation von Grauwertobjekten und ihre Implementierung durch den Processor GIPP. J. of Information Processing and Cybernetics EIK 21(1985) 7/8, 343-354.

/2/ Roesler, U., G. Schwarze , K. Troeger: Digitale Bildverarbeitung und -auswertung fuer die Bruchkontrolle von Schneidwerkzeugen. Bild und Ton 38(1985) 12, 375-378

List of the used abbreviations

data lines

DI	(picture) data input
PD	programming data
SEL	select signals for register programming
WR	write signal for register programming
RD	read signal for register programming
RX	internal register contents
DT	clock signal for DI
AKT	control signal for high speed mode
BER	enable signal
ET,T8,T7,T6	diverse clock signals
RI, RI3, RI5	diverse control signals
BA	mode select signal
EZS	external counter input
ZE	line end signalisation
AZ	adress-connector to line memory
ZP	output line position
T	clock signal for computation
TE	clock enable
TSV	data connector to table memory
TM	table contents
AT	address connector to table memory
DO	(picture) data output
GUL	data valid signalisation
DC	internal strobe value
Z	central pixel value
E	result register contents
V	value for comparison
MA	mask recognition signal
TCL	diverse clock and control signals

function blocks

RCU	register/comparator unit
DEC/RU	decoder/register unit
DIC	data input controller
CSR	clock shift register
M-FF	mask flip-flop
CC	comparator controller
RR/RRC	result register/controller
ZEC/PAC	line end, position and address controller
TS	table selector
SU	subtraction unit

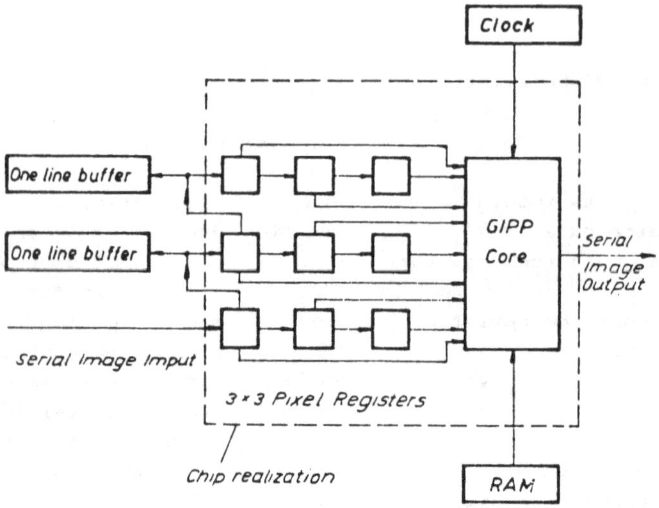

Fig.1 Block diagram of the processor GIPP

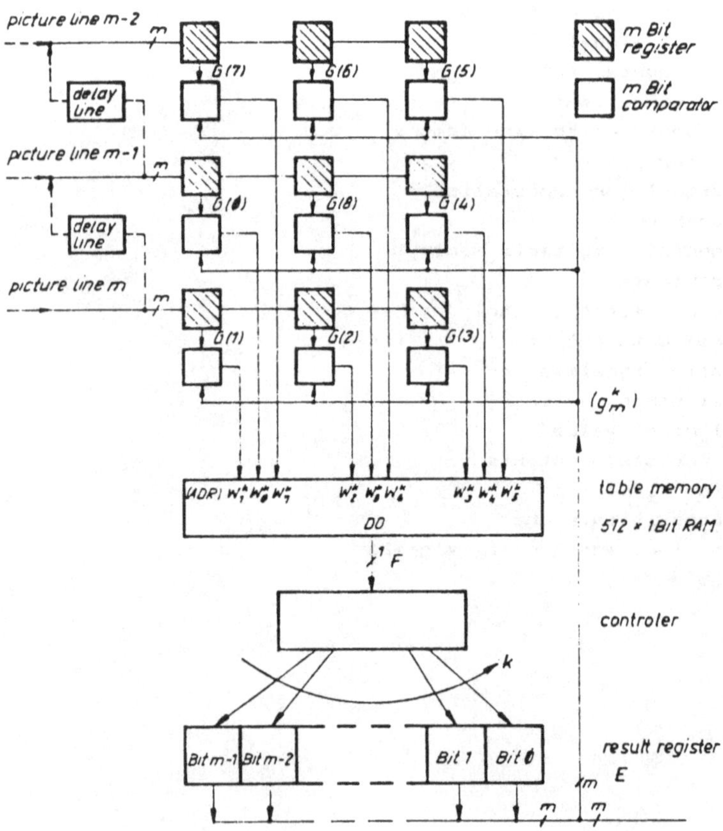

Fig.2 Fundamental hardware structure of a GIPP

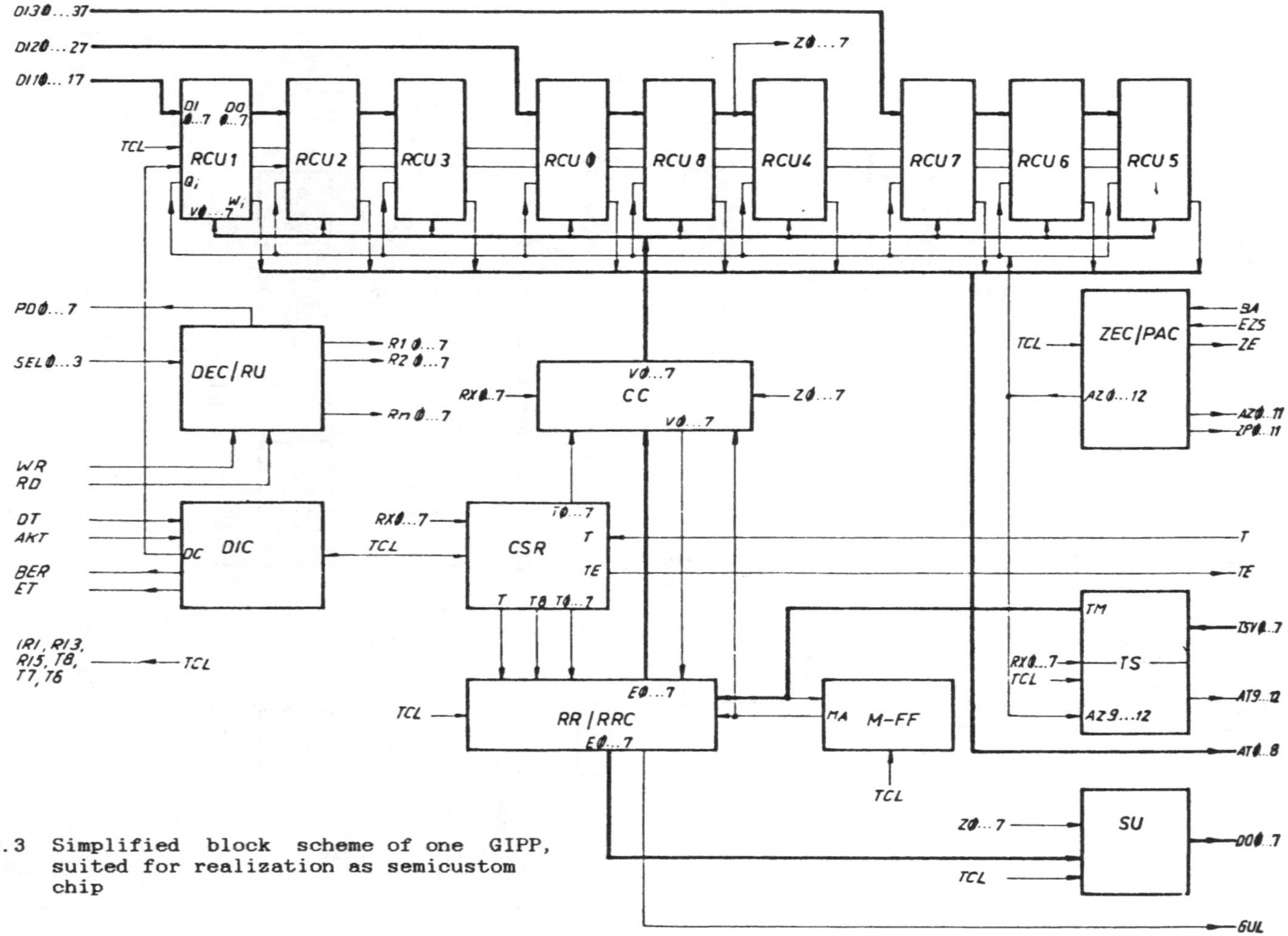

Fig.3 Simplified block scheme of one GIPP, suited for realization as semicustom chip

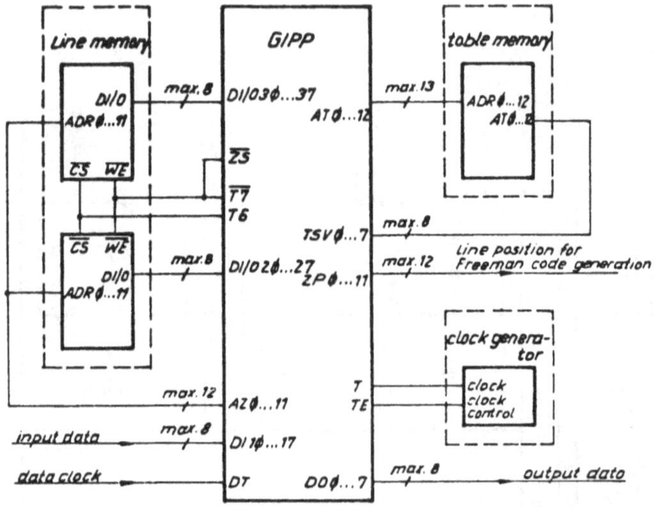

Fig.4 General connection scheme of the main peripheral elements to the GIPP

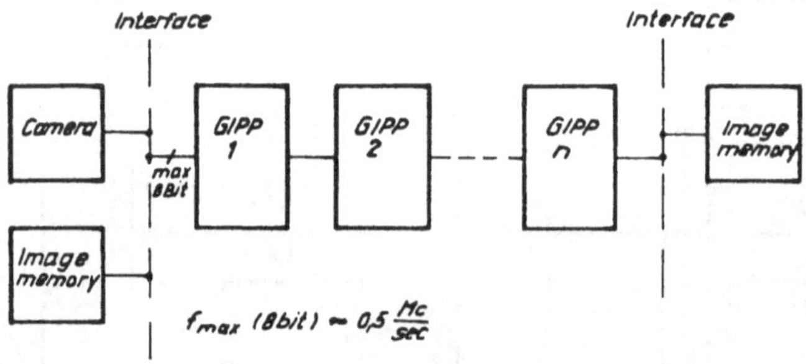

Fig.5 Pipeline of n GIPP's for simultaneous running of n different operators

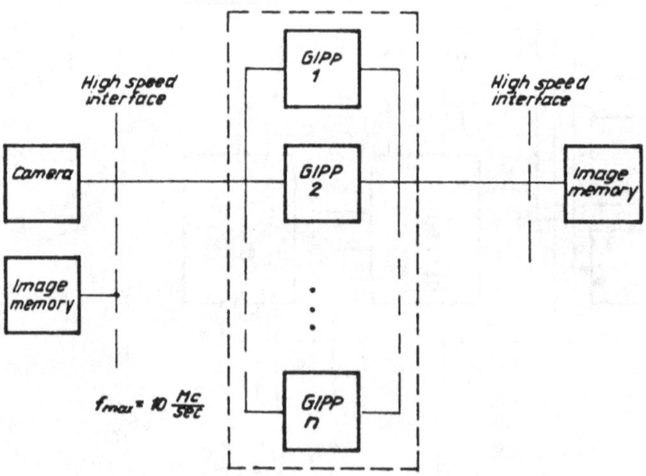

Fig.6 Cyclic arrangement of m GIPP's for accelerated processing

DESIGN OF A HIGH-PERFORMANCE BINARY IMAGE PROCESSOR

J. Owczarczyk[1)]

Abstract

Binary image analysis is one of the most useful techniques available in many applications e.g. robot vision research, biomedical signal processing. However, a real-time analysis of binary images imposes very high performance requirements.

In this paper the systematic design of a very fast processor called TRIP is described in detail. TRIP (Table-driven Row-oriented Image processor) is a SIMD machine with 256 processing elements operating in parallel on separate one-bit wide data streams. TRIP processes a two-dimensional binary image row by row. A computational model for such a processor is a cellular one-dimensional array (cellular string).

The TRIP architecture has been optimized for real-time analysis of binary images. Due to the design strategy used and efficient implementation of cellular logic operators its processing througput exceeds one bilion (10^9) local parallel operations per second. TRIP can accomplish a wide range of complex real-time image analysis tasks using dynamically rewritable microprogram architecture.

1. Introduction

The advent of VLSI technology makes parallel processors more and more attractive as a practical way to increase effective computing power without inordinate increase in cost. Examination of common computations in digital image processing shows that even with current algorithms there is an enormous amount of parallelism to be exploied.

In computerized image processing a majority of computations is of local nature i.e. a new value of a picture element (pixel) is a function

1) Institute of Computer Science
Polish Academy of Sciences
00-901 Warsaw, PKiN, PO Box 22, Poland

of the pxel original value and of its neighbouring pixel values. In the literature this local computation is often refered to as local operation, neighbourhood operation, or cellular logic operation. A number of different architectures has been proposed within the image processing area to fully exploit the possibility of the simultaneous (parallel) computation of local operations. Overviews of this class of systems have been given by Preston [7] and by Danielsson and Levialdi [3] .

In this paper the systematic design of a high-performance parallel processor for binary image analysis is presented. Its architecture, the implementation method of local operators (and hence its processing element structure) differ from the current SIMD image processors such as, for example, CLIP 4 [8] . To explain fully the motivations for the design strategy we discuss at first some computational aspects of common algorithms for binary image analysis and processing.

2. Computational aspects of binary image processing

This section is concerned with a description of fundamental operations which are performed during various tasks for binary image analysis and processing. We recall that in a binary image the pixels are either 0-elements or 1-elements. Usually 0-elements constitute the background whereas 1-elements create the picture.

The simplest operation on binary images is a point-by-point Boolean operation. The Boolean operations between two binary images are widely used for combination partial results in various binary image analysis tasks. This class of operations may be implemented in hardware by a two-input Boolean processor.

An operator L is local parallel operator if it can be represented by:

$$\tilde{a}_{ij} = f(\, a_{ij}, \{a_{kl}\} \mid a_{kl} \in E(i,j)) \qquad (1)$$

where $A=\{a_{ij}\}$ is an input image, $\tilde{A}=\{\tilde{a}_{ij}\}$ is an output image, f is a nine-operand binary function, and E(i,j) is the 8-neighbourhood of the point (i,j).

Local parallel operations cause the simultaneous change of the state of every pixel as a function of the state of its nearest neighbours.

There is a number of binary image processing tasks which can be realized by means of a single local parallel operator e.g. contour extraction, smoothing (majority voting, pepper-and-salt noise removal), erosion (shrink), dilation (expand). They are discussed in detail in [6]. In some algorithms a local parallel operator may be applied iteratively until invariant output is reached. In others, a sequence of different local parallel operators is repeated as for example in the Arcelli et al method of thinning [1] where eight operators are used. Yet another example is a filtering process by means of openings and closings [4,10].

From a series of local parallel operators combined with Boolean operations between partial results obtained during the process complex binary image analysis tasks may be composed as e.g. the skeletonization method proposed in [2].

Some computations are data dependent i.e. the number of operations applied depends on the processed image structure. To terminate such a computation a mechanism must be included in the processor design. It can be an OR gate which is connected to all bits of the result bitplane produced by the last operation or, alternatively, it may be a "no change" flag which indicates whether or not there is a change between the result and operand bitplane.

In binary image processing a local parallel operator can be regarded as a nine-input Boolean function. A hardware implementation of such a function requires a Boolean processor with 512 ($=2^9$) input states and two output states. The total number F of possible functions is therefore given by $F = 2^{2^9}$. If the Boolean processor is to execute all functions then 512 ($=\log_2 F$) control lines should be used to specify its internal logic. This number cPis intolerably high with respect to practical realisation of the wiring scheme. To keep the number of control lines at a reasonable level in practical array processors only a subset of possible functions is directly implemented in hardware. Clearly, the processing element (PE) logic structure reflects the chosen subset. It should be pointed out, however, that there are many function subsets that can be implemented but it is difficult to predict which ones will be useful enough to justify the PE complexity [9].

Alternatively, any local parallel operator can be considered as a mapping from nine input bits into one output bit. Such a mapping can be implemented in hardware by means of a suitably filled table stored in a RAM module. The assembled operands form a nine-bit address and the result is produced by the table look-up. The look-up-table (table-driven) implementation of local parallel operators is very fast provided that a multiplexed and overlapped scheme of table loading is used [5].

3. Table-driven row-oriented image processor

TRIP (Table-driven Row-oriented Image Processor) is a SIMD machine with 256 processing elements operating in parallel on separate one-bit wide data streams under the control of a single control unit. TRIP processes a two-dimensional binary image row by row. A computational model for such a processor is a cellular one-dimensional array (cellular string). Such a row-oriented architecture provides a compromise between purely sequential cellular logic processors, which are to slow for real-time applications, and purely 2D array processors, which require too much hardware for large arrays. The TRIP processing throughput exceeds 10^9 local parallel operations per second [5].

TRIP is a modular system. The overall hardware, including the image memory and processor array, is divided into 16 identical segments and each of them contains 16 PEs.

Fig. 1 ilustrates the basic structure of the TRIP processing element. It consists of four main components: the local memory (32 columns of the image memory), the cellular logic (CL) unit, the Boolean processor, and the A register.

The TRIP image memory consists of 32 bitplanes (each of 256 x 256 bits) which is equivalent to the PE local memory capacity of 32 bits. As compared to local memory sizes in most current array processors (intended mainly for gray-level image processing), this memory capacity is smaller. However, TRIP was intended for binary images and its PE memory size matches that of CLIP 4 image processor which enables utilization of the binary image processing algorithms which have been already developed for CLIP 4.

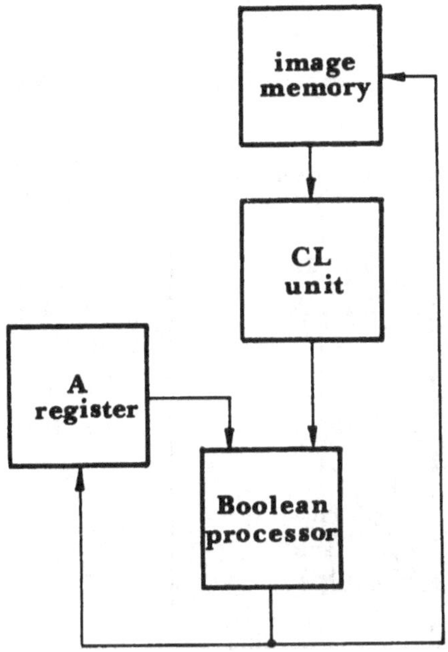

Fig. 1. Block diagram of the TRIP processing element

The CL unit is used for performing local parallel operators. To obtain the operand values three consecutive rows of the processed image are stored at the same time in the cascade of delay registers as shown in Fig. 2. The 9-bit binary data of every neighbourhood are assembled and used as addresses for table look-up in a 512 entry table stored in a RAM module. The contents of the table determine the nature of the operation applied. The neigbourhood interconnection network is used to distribute the pixels into corresponding neighbourhoods.

The main problem of table-driven image processors i.e. overhead required for table loading (a table of 512 adresses must be loaded prior to each new local parallel operation) has been overcome in TRIP by using a pair of table memories and by multiplexed and overlapped scheme of their loading. While one of them is used in an operation the second at the same time is loaded for the next local operation as suppplied by a microcode word. After each operation their roles are reversed. The pair of look-up tables is stored in a 1024x1 RAM with the highest-order address bit used for their selection.

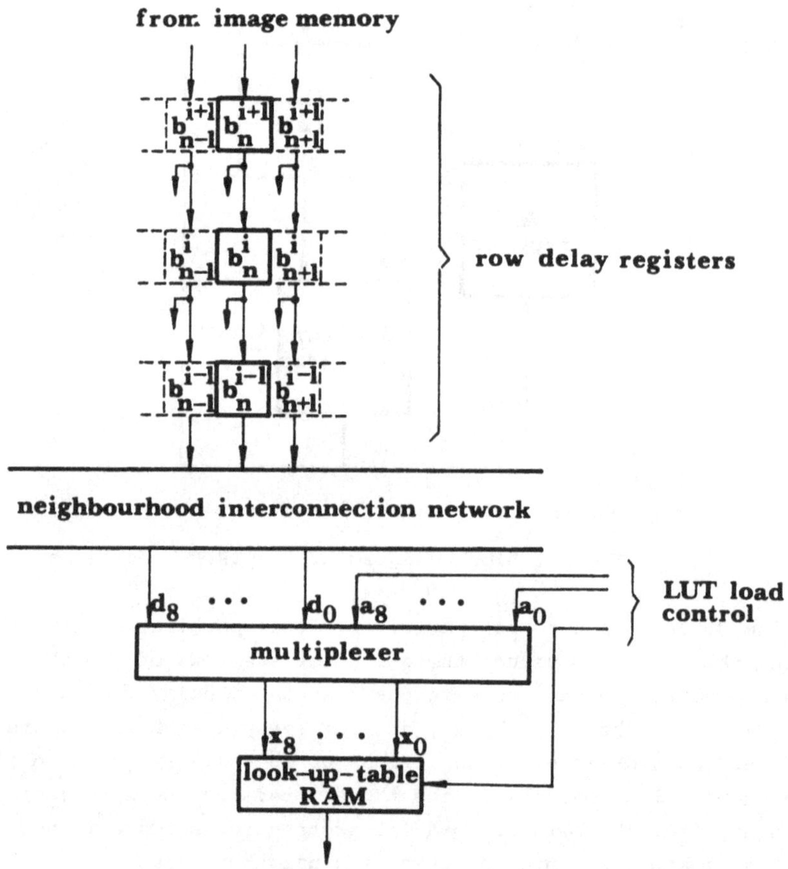

Fig. 2. Simplified logic diagram of the CL unit

The high-speed performance precludes computation of the table contents at run time. Therefore, a number of precomputed tables must be stored in a fast ROM memory. Tn the TRIP microcode word a provision has been made for a set of 256 tables which may be used in computations. The proposed scheme of implementation of local parallel operations is very flexible. By exchanging the ROM chip various function subsets can be tested and used without affecting the hardware of processing elements.

The A register is a working store with the size of one column. It is used as an accumulator. The A register is not directly connected to the image memory (IM), however, it may be loaded from IM through CL unit and the Boolean processor without involving any more memory or clock cycles; when A is loaded from CL, CL is loaded from IM.

The result of a local parallel operation is combined by the Boolean processor with one bit from the A register. The executed result can be stored in the image memory and/or in the A register as supplied by the routing bits in a microcode word. By a series of local parallel operations combined with Boolean operationd between partial results obtained during the process various binary image analysis tasks may be performed.

One important aspect of parallel computation is data access, for even if there are enough processors to exploit the available parallelism, they can only do so if the I/O structure enables suitably fast data transmission. The I/O structure of parallel processor may be implemented in a number of different ways, however, to achieve satisfactory performance data access must be row-parallel. This means that during one I/O cycle a whole row written/read in parallel to/from the image memory.

In TRIP the image data are input to and output from the image memory in 16-bit words via I/O buffer registers in time-multiplexed mode. The I/O operations are partially overlapped with processing. As I/O transfers use asynchronously only the segment buffer registers, processing only needs to be interrupted for two clock cycles (one for input and one for output) per row.

Two mechanisms are included in TRIP for global information extraction. The first is a counter of number of bits set in the result bitplane. The bit-counting operation is a very poweful feature extraction scheme with many applications. The second mechanism is a "no change" flag which indicates any change between the result and operand bitplane. This is most frequently used in data dependent operations for program control e.g. to determine when a loop has terminated.

TRIP was intended primarily for research on binary images and its cost was of main concern so as to make it available for research use. It may be implemented with commercially available integrated circuits. Optimized for use on two-dimensionally structured data as encountered in digital image processing it provides high throughput and offers great flexibility.

References

[1] Arcelli, C., Cordella, L., Levialdi, S.: Parallel thinning of binary pictures. Electronics Letters, 11 (1975), 149-149.

[2] Arcelli, C., Cordella, L., Levialdi, S.: From local maxima to connected skeletons. IEEE Trans. PAMI-3 (1981) 2, 134-143.

[3] Danielsson, P., Levialdi, S.: Computer architecture for pictorial information systems. Computer, 14 (1981) 11, 53-67.

[4] Goetcherian, V.: From binary to grey tone image processing using fuzzy logic concepts. Pattern Recognition, 12 (1980) 7-15.

[5] Owczarczyk, J.: Parallel processor for real-time analysis of binary images. Electronics Letters, 22 (1986) 1179-1181.

[6] Preston, K. et al: Basics of cellular logic with some applications in medical image processing. Proceedings of the IEEE, 67 (1979) 826-856.

[7] Preston, K.: Cellular logic computers for pattern recognition. Computer, 16 (1983) 1, 36-47.

[8] Preston, K., Duff, M.: Modern cellular automata: theory and applications. PLenum Press, London 1984.

[9] Reeves, A.: A systematically designed binary array processor. IEEE Trans. C-29 (1980) 278-287.

[10] Serra, J.: Image analysis and mathematical morphology. Academic Press, London 1982.

Band 1

Inverse and Improperly Posed Problems in Differential Equations

Proceedings of the Conference on Mathematical and Numerical Methods, held in Halle/Saale (GDR) from May 29 to June 2, 1979

Edited by Gottfried Anger
296 Seiten, 1979 (vergriffen)

Band 2

Fundamentals of Computation Theory FCT '79

Proceedings of the Conference on Algebraic, Arithmetic and Categorial Methods in Computation Theory, held in Berlin/Wendisch-Rietz (GDR) September 17–21, 1979

Edited by Lothar Budach
576 Seiten, 1979

Band 3

Decidability and Generalized Quantifiers

By Andreas Baudisch, Detlef Seese, Hans-Peter Tuschik, and Martin Weese
XII, 235 Seiten, 1980

Band 4

Zur Elastizitäts- und Thermoelastizitätstheorie

Potentialtheoretische Methoden in Anwendung auf verschiedene Modelle

Von Lothar Jentsch und Johannes Maul
157 Seiten, 1980

Band 5

Systems Analysis and Simulation 1980

Proceedings of the International Symposium held at Berlin (GDR) September 1–5, 1980

Edited by Achim Sydow
496 Seiten, 1980 (vergriffen)

Band 6

Nichtlineare diskrete Optimierung

von Eberhard Girlich und Michail M. Kowaljow
218 Seiten, 1981

Band 7

Completeness Problems in the Structural Theory of Automata

By Jürgen Dassow
148 Seiten, 1981

Band 8

Darstellung von Lösungen linearer elliptischer Differentialgleichungen

Von Günther Wildenhain
92 Seiten, 1981

Band 9

Introduction to the Spectral Theory of Operators in Spaces with an Indefinite Metric

By I. S. Iohvidov, M. G. Krein, and H. Langer
120 Seiten, 1982 (vergriffen)

Band 10

Nonlinear Time-discrete Systems

A General Approach by Nonlinear Superposition
By Michael Gössel
115 Seiten, 1982 (vergriffen)

Band 11

Preprimal Algebras

By Klaus Denecke
162 Seiten, 1982

Band 12

Ergodic Theory and Related Topics

Proceedings of the Conference held in Vitte/Hiddensee (GDR) October 19–23, 1981

Edited by Horst Michel
232 Seiten, 1982

Band 13

Singularly Perturbed Differential Equations

By H. Goering, A. Felgenhauer, G. Lube, H.-G. Roos, and L. Tobiska
176 Seiten, 1983

Band 14

Leben wir in einer Volterra-Welt?

Ein ökologischer Zugang zur angewandten Systemanalyse

Von Manfred Peschel und Werner Mende
233 Seiten, 1983

Band 15

Markov Decision Problems with Countable State Spaces

Optimality Criteria – Algorithms – Clustering

By Hans Michael Dietz and Volker Nollau
174 Seiten, 1983

Band 16

Stochastische Verfahren zur Suche nach dem Minimum einer Funktion

Von Ryszard Zieliński und Peter Neumann
133 Seiten, 1983

Band 17

Invertibility and Asymptotics of Toeplitz Matrices

By Albrecht Böttcher and Bernd Silbermann
200 Seiten, 1983

Band 18

Structural Induction on Partial Algebras

Introduction to Theory and Application of Partial Algebras – Part II

By Horst Reichel
205 pages, 1984

Band 19

Algebraic Methods for Toeplitz-like Matrices and Operators

By Georg Heinig and Karla Rost
212 pages, 1984

Band 20

Frege Conference 1984

Proceedings of the International Conference held at Schwerin (GDR), September 10–14, 1984

Edited by Gerd Wechsung

408 pages, 1984

Band 21

Numerical Analysis of Selected Semilinear Differential Equations

Edited by Christian Großmann, Thomas Riedrich, and Hartmut Schönheinz

203 pages, 1984

Band 22

Singular Ordinary Differential Operators and Pseudodifferential Equations

By Johannes Elschner

200 pages, 1985

Band 23

Lotka-Volterra-Approach to Cooperation and Competition in Dynamic Systems

Proceedings of the 5th Meeting of UNESCO's Working Group on Systems Theory held on the Wartburg, Eisenach (GDR) at March 5–9, 1984

Edited by Werner Ebeling and Manfred Peschel

271 pages, 1985

Band 24

Convergence Structures 1984

Proceedings of the Conference on Convergence held in Bechyně (Czechoslovakia), September 24–28, 1984

Edited by Josef Novák, Werner Gähler, Horst Herrlich, and Jan Mikusiński

254 pages, 1985

Band 25

Parcella '84

Proceedings of the II. International Workshop on Parallel Processing by Cellular Automata and Arrays held in Berlin (GDR), September 25–27, 1984

Edited by Wolfgang Händler, Tamás Legendi, and Gottfried Wolf

239 pages, 1985

Band 26

Multiobjective and Stochastic Optimization Based on Parametric Optimization

By Jürgen Guddat, Francisco Guerra Vasquez, Klaus Tammer, and Klaus Wendler

175 pages, 1985

Band 27/28

Systems Analysis and Simulation 1985
I: Theory and Foundations
II: Applications

Proceedings of the International Symposium held in Berlin (GDR), August 26–31, 1985

Edited by Achim Sydow, Manfred Thoma, and Robert Vichnevetsky

per volume 495 pages, 1985

Band 29

Parcella '86

Proceedings of the III. International Workshop on Parallel Processing by Cellular Automata and Arrays held in Berlin (GDR), September 9–11, 1986

Edited by Tamás Legendi, Dennis Parkinson, Roland Vollmar, and Gottfried Wolf

272 pages, 1986

Band 30

Fuzzy Sets Applications, Methodological Approaches, and Results

Proceedings of the International Workshop on Fuzzy Sets Applications cosponsored by IIASA and convened in cooperation with IAEA held on the Wartburg, Eisenach (GDR), at March 3–8, 1985

Edited by Steffen Bocklisch, Sergei Orlovski, Manfred Peschel, and Yasushi Nishiwaki

233 pages, 1986

Band 31

Mathematical Methods of Specification and Synthesis of Software Systems '85

Proceedings of the International Spring School, held in Wendisch-Rietz (GDR), April 22–26, 1985

Edited by Wolfgang Bibel and Klaus P. Jantke

245 pages, 1986

Band 32

A Model Theoretic Oriented Approach to Partial Algebras

Introduction to Theory and Application of Partial Algebras – Part I

By Peter Burmeister

319 pages, 1986

Band 33

Finite-difference Methods on Irregular Networks

A Generalized Approach to Elliptic Problems of Second Order

By Bernd Heinrich

206 pages, 1986

Band 34

Modern Algorithms for Large Sparse Eigenvalue Problems

By Arnd Meyer

125 pages, 1987

Band 35

Parametric Optimization and Related Topics

Edited by Jürgen Guddat, Hubertus Th. Jongen, Bernd Kummer, and František Nožička

407 pages, 1987

Band 36

Discretization in Differential Equations and Enclosures

Edited by Ernst Adams, Rainer Ansorge, Christian Großmann, and Hans-Görg Roos

250 pages, 1987

Band 37

Factorization of Measurable Matrix Functions

By Georgi Semenovich Litvinchuk and
Ilya Matveyevich Spitkovskiĭ

372 pages, 1987

Band 38

Parallel Algorithms and Architectures

Proceedings of the International Workshop
on Parallel Algorithms and Architectures
held in Suhl (GDR), May 25–30, 1987

Edited by Andreas Albrecht, Hermann Jung,
and Kurt Mehlhorn

205 pages, 1987

Band 39

Parametric Integer Optimization

By Bernd Bank and Reinhard Mandel

138 pages, 1987

Band 40

Computer Analysis of Images and Patterns 1987

Proceedings of the II. International Conference
CAIP '87 on Automatic Image Processing,
held in Wismar (GDR), September 2–4, 1987

Edited by Leonid P. Jaroslavskiĭ,
Azriel Rosenfeld, and Wolfgang Wilhelmi

282 pages, 1987

The processing of pictorial signals and the automatic analysis of images has developed into one of the main application fields of information processing and into an active research area of computer science.
They have a strong influence on new generations of computers with the ability of understanding natural scenes and the automation of complicated inspection and manipulation procedures.
Image processing and image pattern recognition contribute to the introduction of novel information processing principles (especially parallel ones) and the invention of corresponding algorithms as well. On the other hand methods and tools of Artificial Intelligence are being utilized for image analysis tasks more and more.
Finally we observe that an own theoretical framework is being established by adoption and enrichment of geometry and topology jointly and concurrently with Computer Graphics.
Reader will find full length papers of selected contributions to the international conference CAIP '87 concerning the above mentioned topics. The volume contains surveys and reports on detailed results as well.

Die Verarbeitung von Bildsignalen und die automatische Auswertung von Bildern haben sich zu einem wichtigen Anwendungsgebiet der Informationsverarbeitung und einem sehr aktiven Feld der Informatikforschung entwickelt.
Von ihnen gehen nicht nur starke Impulse für neue Generationen von Rechnern mit der Fähigkeit zum Verstehen natürlicher Szenen und für die Automatisierung komplizierter Kontroll- und Handhabevorgänge aus.
Bildverabeitung und Bilderkennung tragen zur Einführung neuartiger, insbesondere paralleler, Verarbeitungsprinzipien und zur Erfindung entsprechender Algorithmen bei. Andererseits werden zunehmend Methode und Werkzeuge der Künstlichen Intelligenz für Bildanalyseaufgaben nutzbar gemacht.
Schließlich entstehen in Kooperation und im Wettbewerb mit der Computergraphik eigene theoretische Grundlagen durch Aneignung und Bereicherung von Geometrie und Topologie.
Zu all diesen Punkten wird der Leser durch Langfassungen ausgewählter Beiträge zur Konferenz CAIP '87 informiert.
Der Band enthält sowohl Übersichten als auch Berichte über aktuelle Resultate.

ISBN 3-05-500451-5
ISSN 0138-3019

MIX
Papier | Fördert gute Waldnutzung
FSC® C083411

Zeitfracht Medien GmbH
Ferdinand-Jühlke-Straße 7
99095 Erfurt, Deutschland
produktsicherheit@kolibri360.de